AF333545

THERMO
LOOP APPLICATIONS
IN
MATERIALS SYSTEMS
DYNAMIC

FIRST EDITION:

DONALD L. JOHNSON

Professor Emeritus of Mechanical Engineering,
Metallurgy Program, University of Nebraska-Lincoln,
Lincoln, Nebraska

GLENN B. STRACHER

Professor of Geology, East Georgia College,
University System of Georgia,
Swainsboro, Georgia

A Publication of The Minerals, Metals & Materials Society
420 Commonwealth Drive
Warrendale, Pennsylvania 15086
(412) 776-9000

Printed in the United States of America
Library of Congress Catalog Number 94-073747
ISBN Number 0-87339-270-1

© 1995

If you are interested in purchasing a copy of this book, or if you would like to receive the latest TMS publications catalog, please telephone 1-800-759-4867.

CONTENTS

1 INTRODUCTION

**2 THE FIRST LAW OF THERMODYNAMICS
AND ENTHALPY**

PREFACE

This book is intended for use in an initial one-semester course in "materials" thermodynamics, although it could be used for self-study or as a supplement to a text selected for a two-semester undergraduate-graduate sequence. Each chapter is designed so that the student can learn to use thermodynamics as a problem solving tool for a broad range of materials applications. The book is written from notes developed in a course entitled "Thermodynamics of Alloys." Although this course was originally designed as an elective in mechanical engineering at the University of Nebraska-Lincoln, the broader perspective of thermodynamics of materials has emerged as a result of student contacts in earth science, mechanical, chemical, and corrosion engineering, and electrical materials.

For many scientists and engineers, thermodynamics may be most useful for applications where quickly obtained numerical estimates are all that is needed. Such a situation may prevail for those working in materials-related areas where the generation of thermodynamic data is not an objective in itself. The earth scientist, for example, may need to interpret the stability of mineral assemblages in a rock by using thermodynamic data available in the literature. On the other hand, metallurgists and chemists as well as mechanical, electrical, and materials engineers may be concerned with elevated temperature phase equilibria, phase transformations, and environmental reactions. For the corrosion and chemical engineer, the concept of potential (EMF) is very important in corrosion control as it relates to the Nernst Equation and associated "Pourbaix" diagrams.

Fundamentals of classical thermodynamics are briefly reviewed and systematically introduced into problem solving using "thermodynamic loop" or TL analysis. This concept, derived from Kirchhoff's law, is analogous to cyclic analysis used to evaluate engine performance in mechanical systems. Virtually all material thermodynamic applications can be analyzed by TL analysis. TL analysis is a method of organization whereby the known state of a system is combined with thermodynamic properties of materials obtainable from the literature to calculate system properties in another state. TL analysis is a powerful analytical tool because it divides a problem into parts and graphically structures it in such a manner as to provide the student with a clearly depicted solution path. In addition, TL analysis eliminates some of the need for memorizing detailed formulas because it can be used to derive them. Extensive use of TL analysis is a unique feature of this book.

The first chapter introduces thermodynamics as a science and defines important terms. The thermodynamic loop is introduced in the second chapter. Example applications in the second chapter include constant and variable temperature enthalpy calculations and supercooling. The Second and Third Laws of Thermodynamics are introduced in the third chapter and the concept of entropy as a stability criterion is discussed. Example applications include constant and variable temperature calculations and supercooling. Finally, thermodynamic efficiency is defined graphically from a temperature-entropy (T-S) diagram. The fourth chapter is devoted to Gibbs free energy, stability criteria, and introduction of the Ellingham diagram. Example TL applications include variable temperature and pressure equilibrium calculations for closed systems. The chapter includes the application of free energy stability criteria to the same supercooling problems presented in Chapters 2 and 3. A comparison with entropy stability criteria illustrates the simplicity of Gibbs free energy calculations in analyzing reactions at equilibrium. Chapter 4 concludes with numerous thermodynamic relations that are used later in the book. The fifth and sixth chapters further develop the analysis of systems at equilibrium. Chapter 5 covers systems open with respect to gaseous phases (variable gas composition), applications of the Ellingham diagram, and illustrations of techniques for calculating gas phase equilibria over pure condensed phases. These techniques are expanded in Chapter 6 to include variable condensed phase compositions. Chapter 6 begins by defining fugacity, thermodynamic activity, and partial properties in terms of volume and then builds upon these concepts to define other common mixing properties. Discussion is limited to two-component systems. Partial properties are the basis upon which ideal, dilute, and regular solution models are subsequently derived. The chapter concludes with several problems including alloy oxidation, application of the Gibbs-Duhem equation, and development of Sievert's law. Chapter 7 is devoted to the analysis of binary phase diagrams. There is a wealth of untapped information in phase diagrams and the intent of this chapter is to illustrate the utility of the TL concept in extracting, at the very least, useful estimates of partial molar properties from liquidus, solidus, and solvus equilibria. These estimates are incorporated into such applications as prediction of purity, carburizing potential, and analysis of reactions. In addition, techniques are developed for estimating the effect of pressure on phase boundaries. Finally, phase rule concepts are introduced. Chapter 8 introduces the concept of the equilibrium constant and illustrates its utility in more complex three component systems. The equilibrium constant, K_{eq}, is generally not required to solve any of the problems presented in this book. Experience has shown that the student or casual user often employs it improperly. In this connection, reliance upon computer programs to solve thermodynamic problems is left to the reader who has mastered basic concepts. The chapter concludes with a discussion of corrosion cell polarity.

Numerous people helped the authors during the development and preparation of the final draft and its accompanying solutions manual. Sincere appreciation goes to Dr. Russell C. Nelson, Department of Mechanical Engineering, University of Nebraska-Lincoln, for his review and critique of manuscript drafts. Thanks also goes to Dr. Dana J. Medlin, Department of Metallurgical and Materials Engineering, Colorado School of Mines, for helpful comments on the initial draft. The authors are especially grateful to Janet L. Stracher for valuable and untiring assistance with the preliminary preparation of figures and graphs and for final editing of the solutions manual.

We also thank Bob Makowski, Janet Urbas, Dane Semonian, and the staff at The Minerals, Metals and Materials Society of the American Institute of Mining, Metallurgical and Petroleum Engineers for their guidance through the entire project. Finally, the authors wish to thank their families for their patience and encouragement during the many long hours devoted to writing this book.

DONALD L. JOHNSON
GLENN B. STRACHER

SYSTEMS AND UNITS

The standardization of physical quantities is necessary so that measurements based on these quantities or combinations of them mean the same thing in various disciplines. Quantities chosen as *standards* represent the *basic units* of measurement. Basic units are selected so as to represent the smallest number of physical quantities from which all others may be derived. *Derived units* then, are simply combinations of the basic units. Because of the variety of units used, conversions are presented in this section.

The Fourteenth Conference on Weights and Measures in 1971 adopted three classes of units internationally accepted today: (1) base units in Table SI-1, (2) derived units in Table SI-2, and (3) supplementary units (NBS Publication 330, 1972). The symbols for derived units are obtained by expressing the base and/or supplementary units with the mathematical symbols for division and multiplication. For example, the derived SI unit for force is $kg \cdot m/s^2$ (to which is assigned the name Newton, symbolized by a capital N). Likewise, the derived SI unit for power is $[(kg \cdot m/s^2) \cdot m]/s = (Newton \cdot m)/s = joule/s$ (where joule is symbolized by a capital J).

The SI system excludes using units such as the dyne (1 dyne = 10^{-5} N) and the erg (1 erg = 10^{-7} J), disapproves of using the torr [1 torr = 1 mm Hg = (101.325/760) kPa] and the calorie (1 thermochemical calorie = 4.184 J), and tolerates using the bar (1 bar = 10^5 Pa) and the standard atmosphere (1 atm = 101.325 kPa). Because of past and currently widespread use of the calorie, bar, and standard atmosphere by engineers and scientists, these units will be used on occasion throughout this book.

U.S. Customary Units

In addition to the SI system of units, the U.S. customary or British system of units is commonly used today in the United States, Great Britain, and elsewhere. Conversion factors between SI and U.S. customary units, given in Table CF-1, may be useful.

Table SI-1: SI (Systéme International) Base Units

Quantity	Name	Symbol
Length	meter	m
Mass	kilogram	kg
Time	second	s
Thermodynamic temperature	kelvin	K
Amount of substance	mole	mol
Electric current	ampere	I

Table SI-2: Derived SI Units

Quantity	SI Formula	Special Symbol	Special Name
Area	m^2	A	—
Volume (solids)	m^3	V	—
Volume (liquids)	$10^{-3}\ m^3$	L	liter
Density	kg/m^3	—	—
Force	$kg{\cdot}m/s^2$	N	Newton
Pressure, stress	N/m^2	Pa	pascal
Pressure	101.325 kPa	atm	atmosphere
Pressure	$10^5\ Pa$	—	bar
Work, energy	$N{\cdot}m$	J	joule
Power	J/s	W	watt
Linear velocity	m/s	—	—
Linear acceleration	m/s^2	—	—
Electric charge	$I{\cdot}s$	C	coulomb
Electromotive force	W/I	V	volt

Table CF-1: SI and U.S. Customary Equivalents

Quantity	SI Unit	U.S. Customary Equivalent
Length	25.40 mm	in
Length	0.3048 m	ft
Area	$645.2\ mm^2$	in^2
Area	$0.0929\ m^2$	ft^2
Volume	$16.39\ cm^3$	in^3
Volume	$0.02832\ m^3$	ft^3
Volume liquid	0.9464 L	qt (quart)
Volume liquid	3.785 L	gal (gallon)
Mass	0.4536 kg	lb mass
Mass	14.59 kg	slug
Mass	907.2 kg	ton
Force	4.448 N	lb_f
Force	4.448 kN	kip
Pressure, stress	47.88 Pa	lb_f/ft^2
Pressure, stress	6.895 kPa	lb_f/in^2 (psi)
Pressure, stress	6.895 MPa	$1000\ lb_f/in^2$ (ksi)
Pressure	101,325 Pa = 1 atm	—
Pressure	$10^5\ Pa$ = 1 bar	—
Work, energy	1.356 J	$ft{\cdot}lb_f$
Heat, energy	1054.8 J	BTU (British thermal unit)
Power	1.356 watts	$ft{\cdot}lb_f/s$
Power	745.7 watts	hp

NOTATION

P	Pressure
P_i	Partial Pressure of Gas Component i
V	Molar Volume
v	Specific Volume (mass basis)
l	Length
L	Liter
ρ	Density
T	Temperature
T^{Tr}	Transformation Temperature
T^f	Fusion Temperature, Pure Component
°C	Degrees Centigrade
°F	Degrees Fahrenheit
K	Degrees Kelvin
R	Degrees Rankine or Ideal Gas
J	Joules
kJ	Kilojoules
θ	General State Function
m	Mass
gm	Gram
kg	Kilogram
lb_m	Pound Mass
lb_f	Pound Force
W	Work/Mol or Work/Unit Volume
Q	Heat (Transfer)/Mol
q	Heat (Transfer)/Unit Mass
E_P	Molar Potential Energy
E_K	Molar Kinetic Energy
E_T	Total Energy/Mol
U	Molar Internal Energy
u	Internal Energy (mass basis)
H	Molar Enthalpy
ΔH	Molar Enthalpy Change
ΔH^0	Standard State Molar Enthalpy Change
ΔH^{0f}	Standard State Molar Enthalpy of Formation
ΔH^{Tr}	Molar Heat of Transformation
ΔH^f	Molar Heat of Fusion, Pure Component
ΔH^v	Molar Heat of Vaporization
ΔH^s	Molar Heat of Sublimation
$\Delta H_s, \Delta H_s^0$	Molar or Standard Molar Heat of Solution (Alternately, $\overline{H}_i^m$)
h	Enthalpy (mass basis)
S	Molar Entropy
ΔS	Molar Entropy Change

ΔS^0	Standard State Molar Entropy Change
ΔS^{0f}	Standard State Molar Entropy of Formation
ΔS^f	Molar Entropy of Fusion, Pure Component
$\Delta S_s, \Delta S_s^0$	Molar or Standard Molar Entropy of Solution (Alternately, $\overline{S}_i^m$)
s	Entropy (mass basis)
G	Molar Gibbs Free Energy
ΔG	Molar Gibbs Free Energy Change
ΔG^0	Standard State Molar Gibbs Free Energy Change
ΔG^{0f}	Standard State Molar Gibbs Free Energy of Formation
ΔG^f	Molar Gibbs Free Energy of Fusion
$\Delta G_s, \Delta G_s^0$	Molar or Standard Molar Gibbs Free Energy of Solution (Alternately, $\overline{G}_i^m$)
g	Gibbs Free Energy (mass basis)
A	Molar Helmholtz Free Energy or area
C_p	Constant Pressure Molar Heat Capacity
C_v	Constant Volume Molar Heat Capacity
c_p	Constant Pressure Specific Heat Capacity (mass basis)
c_v	Constant Volume Specific Heat Capacity (mass basis)
a_i	Activity of Component i
γ_i	Activity Coefficient of Component i
Π	Product of Activities
J_a	Activity Quotient
C_i	Concentration of Component i
K_{eq}	Equilibrium Constant
f_i	Fugacity of Component i
μ_i	Chemical Potential of Component i
σ	Uniaxial Stress
ε	Uniaxial Strain
$(s), < >$	Solid: e.g., PbS(s) or <PbS>
$(l), [\ ]$	Liquid: e.g., $H_2O(l)$ or $[H_2O]$
$(g), (\ )$	Gas: e.g., $N_2(g)$ or (N_2)
$'$	Superscript Prime: Used for extensive thermodynamic variables. For example, $\Delta H'$ is the *total* enthalpy change of a substance or a reaction.
$f_{w,i}$	Weight Fraction of Component i
$f_{v,i}$	Volume Fraction of Component i
mole (mol)	Gram Mole, or Gram Atom
pct	Percent
ppm	Parts Per Million (mass basis)
ppb	Parts Per Billion (mass basis)
N_i	Number of Atoms of Component i
N_T	Total Number of Atoms
N_{AV}	Avogadro's Number

k	Boltzmann's Constant (R/N_{AV})
n_i	Number of Moles of Component i
n_T	Total Number of Moles
M	Molar Mass
X_i	Atomic Fraction of Component i in Condensed Phase
Y_i	Atomic Fraction of Component i in Gas Phase
a/o	Atomic Percent
w/o	Weight Percent
h, s	Nonideal Solution Constants
Ω, ω, δ	Nonideal Solution Constants
z	Coordination Number
ϕ, F, Γ	Phases, Degrees of Freedom, and Components respectively: Gibbs Phase Rule
E	Electromotive Force (EMF)
F	Faraday's Constant: Coulomb/equivalent(eq)
n	Valence (eq/mol) or molar mass/equivalent mass where M/n is equivalent mass
α	Volume Thermal Expansion Coefficient or Polymorphic Phase
β	Isothermal Compressibility Coefficient or Polymorphic Phase
η	Thermodynamic Efficiency

SYMBOLS FOR MOLAR PROPERTIES: CHARACTERIZED BY VOLUME*

V_A^0, V_B^0	Molar volume of pure components A and B respectively
$\overline{V}_A, \overline{V}_B$	Partial molar volume of A and B respectively
n_A, n_B	Number of moles of A and B respectively
X_A, X_B	Mole fraction of A and B respectively, where $X_A = n_A/(n_A + n_B)$, $X_B = n_B/(n_A + n_B)$ and $X_A + X_B = 1.0$
$\overline{V}_A^m, \overline{V}_B^m$	Partial molar volume of mixing of A and B respectively, where $\overline{V}_A^m = \overline{V}_A - V_A^0$ and $\overline{V}_B^m = \overline{V}_B - V_B^0$
V^m	Molar volume of mixing where $V^m = X_A \overline{V}_A^m + X_B \overline{V}_B^m$
$\overline{V}_A^{XS}$	Partial molar excess volume of A
$V^{m\prime}$	Total Volume of Mixing where $V^{m\prime} = n_A \overline{V}_A^m + n_B \overline{V}_B^m$

* (1) Identical expressions are used for thermodynamic properties other than volume.

  (2) Ideal properties are designated by "id". For example, $\overline{V}_A^{m,id}$ is the partial molar volume of mixing of A in liquid solution; reference is pure liquid A.

  (3) Phase identification is denoted with a superscript. For example, $\overline{V}_A^{m,l}$ is the partial molar volume of mixing of A in liquid solution; reference is pure liquid A.

TABLE OF PHYSICAL CONSTANTS AND CONVERSION FACTORS

Ideal Gas Constant $\quad R = 8.3144$ joules/(mol·K)

$\qquad = 1.987$ calories/(mol·K)

$\qquad = 82.057$ cm^3·atm/(mol·K)

$\qquad = 0.082057$ liter·atm/(mol·K)

$\qquad = 0.083144$ liter·bar/(mol·K)

$\qquad = 1.987$ BTU/(lb mol·R)

$\qquad = 0.73$ atm·ft^3/(lb mol·R)

Pressure $\quad$ 1 atmosphere (atm) $= 760$ mm Hg

$\qquad = 760$ torr

$\qquad = 14.696$ lb$_f$/in^2

$\qquad = 1.01325$ bar

$\qquad = 101.325$ kN/m^2

$\qquad = 101.325$ kPa

Temperature $\quad$ K $= °C + 273.16$ degrees

$\qquad$ °F $= (9/5)°C + 32.00$ degrees

$\qquad$ R $= °F + 459.67$ degrees

Energy Equivalents $\quad$ 1 calorie $= 4.184$ joules (J)

$\qquad = 4.184 \times 10^7$ ergs

$\qquad = 0.041293$ liter·atm

$\qquad = 41.3223$ cm^3·atm

$\qquad = 3.0855$ ft·lb$_f$

$\qquad = 1.4580 \times 10^{-3}$ atm·ft^3

$\qquad = 3.9683 \times 10^{-3}$ BTU

Avogadro's Number $\quad N_{AV} = 6.0232 \times 10^{23}$/mol

Boltzmann's Constant $\quad k = R/N_{AV}$

Faraday's Constant $\qquad = 95,000$ coulomb/equivalent

INTRODUCTION

1.1 THE SCIENCE OF THERMODYNAMICS

Thermodynamics is concerned with the study of energy transformations and the relationship of these transformations to materials properties. These relationships are derived from *thermodynamic laws*, which are mathematical expressions assumed to hold true for all cases in which the specified conditions of the law are met experimentally. Because its laws are independent of atomic and molecular theory, *classical thermodynamics* can be thought of as a *macroscopic* science.

The purpose of thermodynamics is to predict changes in the properties of some finite portion of space set aside for investigation and to determine the influence, if any, on this portion of space by its surroundings. In the engineering and physical sciences, thermodynamics is used to assess efficiency and predict the occurrence of chemical and physical processes. Unfortunately, thermodynamics generally cannot be used to determine the rate at which a reaction takes place—i.e., it tells what occurs at equilibrium, but not when.

1.2 SYSTEMS, SURROUNDINGS, AND PHASES

Thermodynamic analysis starts with identification of the *system* under study and the location of its boundaries relative to the *surroundings*. For example, materials systems are composed of *phases* in a finite portion of space set aside for investigation. A phase is a physically distinct, mechanically separable portion of a system, e.g., ice in a mixture of ice and water or the mineral olivine in the olivine basaltic rocks of the Hawaiian Islands. Systems composed of one phase are said to be *homogeneous,* while systems composed of two or more phases are said to be *heterogeneous.* The boundaries of a system may be real, such as the walls of a container holding a metal charge, or they may be purely imaginary. Everything outside the system, which either directly influences or has no influence at all on its behavior, constitutes the surroundings. If a system is unaffected by its surroundings it is said to be *isolated.* An isolated system exchanges neither energy nor mass with its surroundings. A *closed system* exchanges only energy with its surroundings whereas an *open system* exchanges both energy and mass with its surroundings.

1.3 MACROSCOPIC STATE OF A SYSTEM

Experimental measurement of thermodynamic properties provides the basis for the complete description of the *macroscopic* state of a system. Hence, the empirical approach is the primary focus of this book. The number of *macroscopic coordinates* or variables needed to define a system depends upon whether the system is open or closed and upon the identity of the boundary between the system and the surroundings. For example, the mineral phases comprising a rock at great depth in the earth's crust may react to form other minerals upon exhumation to the surface. If the exhumed rock is considered the system, the pressure, temperature, and composition recorded under exhumed conditions may be adequate to describe the state of the system at or near the surface. However, if the focus of attention is reaction conditions at some specific point in time at great depth, then an entirely different set of pressure, temperature, and composition data is required. The *macroscopic state of a system* refers to the experimental coordinates needed to define the system in such a manner that it could be duplicated. If the coordinates change in any way, the system is said to undergo a *change in state*. For example, the state of a closed system can be defined by any two of three macroscopic variables: namely, pressure, volume, and temperature. Hence, if one of these is selected as the dependent variable, the other two become the independent variables. If the system undergoes a change in one or more independent variables, it changes state. Microscopic states and their connection to macrostates, briefly discussed in Chapter 3, are the subject matter of statistical thermodynamics.

1.4 EQUILIBRIUM

A system is said to be in a *state of equilibrium* when there is no perceptible change in macroscopic coordinates with time. Such systems *appear* to be at rest. A system in *chemical equilibrium* appears to be at rest but actually involves a balance of reactions opposing one another at equal rates. The equilibrium between a gas and a liquid at the boiling point of the liquid and between a solute and its undissolved supersaturated component are examples of chemical equilibrium.

In addition to chemical equilibrium, systems may exhibit mechanical and thermal equilibrium. A system is said to be in a state of *mechanical equilibrium* when no unbalanced force exists between the internal parts of the system as well as between the system and its surroundings. *Thermal equilibrium* occurs when a system and its surroundings are at the same temperature and no net heat transfer occurs across the boundary.

Any system in chemical, mechanical, and thermal equilibrium is said to be in a state of *thermodynamic equilibrium*. Such systems can be described in terms of time-independent macroscopic coordinates. It should be noted that a system at *steady state* can be described in terms of time-independent macroscopic coordinates, but it is not at equilibrium because the coordinates change if the surroundings are modified. Some equilibrium states can be predicted from everyday observations. For example, it is intuitively known that when a bar of hot metal is quenched in water, the metal and the water reach thermal equilibrium at an intermediate temperature. In more complex systems, the equilibrium state is not predictable from observation, and analytic criteria for equilibrium must be established. Establishing such criteria is important, for example, in order to determine the direction in which a chemical reaction will spontaneously proceed at some instant in time when the macroscopic coordinates of the chemical system are known. A summary of the preceding discussion is presented in Table 1.1.

Table 1.1: Thermodynamic Equilibrium States of A System		
Thermodynamic Equilibrium		
I. Chemical Equilibrium: Balance of opposing reactions at equal rates	*II. Mechanical Equilibrium:* No unbalanced force in the system or between the system and its surroundings	*III. Thermal Equilibrium:* System is at the same temperature as the surroundings

1.5 ADIABATIC AND DIATHERMIC BOUNDARIES

Two types of boundaries, often distinguished in thermodynamics, are *adiabatic** and *diathermic*. Adiabatic boundaries or heat insulators prevent heat transfer between a system and surroundings. Diathermic boundaries or heat conductors allow a system to exchange heat with its surroundings and, in so doing, are themselves influenced during the interaction.

Suppose two or more systems are connected by an adiabatic boundary to each other and by a diathermic boundary to an additional system as in Figure 1.1(a). If this configuration is itself enclosed by an adiabatic boundary, then it can be concluded that the adiabatically connected systems will be in thermal equilibrium with the additional system. If the adiabatic boundary separating the systems is removed and replaced by a diathermic boundary at the same temperature as the original boundary, all systems will be in thermal equilibrium with each other, e.g., Figure 1.1(b). Zemansky and Van Ness (1966, p. 7) expressed these observations as the *Zeroth Law of Thermodynamics*: "Two systems in thermal equilibrium with a third are in thermal equilibrium with each other." The reader is also referred to Thomsen (1962) for an interesting discussion of the Zeroth Law.

---

* Although there is no such thing as a perfect adiabatic heat insulator, experiments may be carried out so as to closely approximate adiabatic conditions for the duration of the experiment.

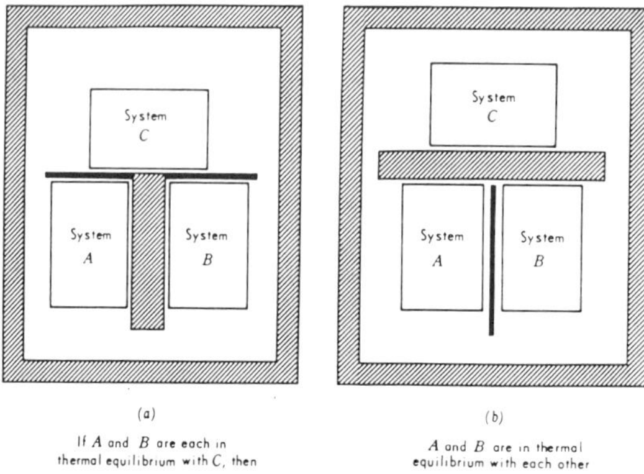

Figure 1.1 Illustration of the Zeroth Law of Thermodynamics. Cross shading indicates adiabatic boundaries; thick dark lines, diathermic boundaries. (From M.W. Zemansky and H.C. Van Ness, 1966, Basic Engineering Thermodynamics, Fig. 1.2. Reprinted by permission of McGraw-Hill, Inc., New York.)

1.6 IRREVERSIBLE AND REVERSIBLE PROCESSES

Changes in state take place by *irreversible* or *reversible processes*. An irreversible process, also known as a *spontaneous* or *natural* process, is one in which a system moves from a nonequilibrium state to one of equilibrium as a consequence of some finite external agent or driving force acting on the system. Although it is natural for all nonequilibrium systems to move toward an equilibrium state, the rate at which the drive towards equilibrium occurs is variable. The freezing of granitic magma intruded into the earth's crust, for example, is a *thermally irreversible* process that can vary from a fraction of a second at the surface of the intrusion to many millions of years deep inside of it. The external agent causing the change in state of the intrusion is the thermal gradient between the magma and the cooler surroundings. An example of *mechanical irreversibility* is the free expansion of a gas from its container into a vacuum as a consequence of differential pressure between the gas and the vacuum. Other examples of spontaneous processes include those displaying *chemical irreversibility*. These include changes in chemical composition (*chemical reactions*), changes in crystal structure but not composition (*polymorphic or allotrophic transformations*), phase changes, and the mixing of substances.

A reversible process, also referred to as a *quasi-static* process, is one in which system properties change due to an infinitesimally small driving force that never produces more than an infinitesimally small displacement from equilibrium. In addition, any change in the system can be reversed by an infinitesimally small change in the driving force. The system can be thought of as passing through a continuum of equilibrium states in such a manner that each infinitesimally small deviation from equilibrium is followed by an equilibrium state.

1.7 PATH INDEPENDENT PROPERTIES, CHANGE IN STATE, AND CYCLIC PROCESSES

It is convenient to express the properties of a system (or its surroundings) in terms of algebraic variables. In the literature, *path independent* properties or variables are referred to by numerous names including: thermodynamic functions, thermodynamic functions of state, thermodynamic variables, thermodynamic properties, state functions, state variables, state properties and point functions.

If θ is a state variable with initial and final values θ_i and θ_f respectively, the change in state of a system with respect to θ is designated as $\Delta\theta = \theta_f - \theta_i$. An infinitesimal change in θ between the initial and final states of the system is designated by the *exact differential* $d\theta$. The symbols Δ and d are used to designate macroscopic or infinitesimal changes respectively, for state functions only.

If the initial state (i) of a system is identical to its final state (f), then a combination of processes that takes the system through a series of changes from i to f is said to be a *cyclic process* or a *cycle*. For any state function θ that has gone through n cycles, where n is an integer > 0,

$$\sum_{a=1}^{n} \Delta\theta_a = 0 \ \text{ or } \ \oint d\theta = 0$$

where $\oint$ designates a *cyclic integral*. An example of a mechanical system that undergoes thousands of cyclic processes per second is the fluid in the power steering pump of an automobile engine. If the engine is properly tuned, the thermodynamic properties of the fluid as it enters the high pressure hose are the same as when it reenters this hose after having first passed through the steering gear box into the return hose and then into the pump reservoir.

State properties are classified as either *extensive* or *intensive*. Extensive properties are characterized by magnitudes dependent on the size of a system or system component. They are often expressed with a superscript prime and include such properties as mass (m), volume (V'), and as will be seen later, internal energy (U'), enthalpy (H'), entropy (S'), Gibbs free energy (G'), and total heat capacity (C'). Intensive properties are macroscopic coordinates characterized by magnitudes independent of the size of a system or a system component. They include pressure (P), temperature (T), stress (σ), strain (ε), surface tension (γ), electric cell Emf (E), specific volume (v), density (ρ), molar heat capacity (C), specific heat capacity (c), volume (V), internal energy (U), enthalpy (H), entropy (S), and Gibbs free energy (G). Intensive properties may be derived from extensive properties by expressing the extensive property on a per unit mass (*specific*), mole, or volume basis. For example, V', the volume of a system component is an extensive thermodynamic property. Dividing volume by the mass or the number of moles, the result is an intensive property such as specific volume ($V'/\text{mass}) = v$ or molar volume ($V'/\text{moles}) = V$.

The path independent nature of thermodynamic properties is illustrated in Figure 1.2. Point 1 (initial state) is defined by P_1, V_1, and T_1, while point 2 (final state) is defined by P_2, V_2, and T_2. Since both points are represented by state properties, the change in these properties is the same for any process or transition path that leads from point 1 to 2. The transition paths, for example, may be characterized by *isothermal ($\Delta T = 0$)*, *isochoric* or *isometric ($\Delta V = 0$)*, or *isobaric ($\Delta P = 0$)* processes. Each path will result in the same values for P_2, V_2, and T_2 at point 2. As illustrated, the paths can be expressed in terms of partial derivatives. Mathematical manipulation of these partial derivatives in conjunction with the appropriate equation of state and thermodynamic relations given in Chapter 4 are useful in computing other desired property changes.

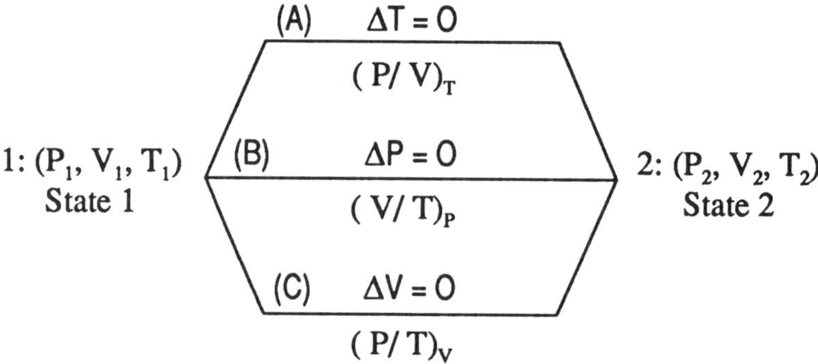

Figure 1.2 Illustration of the path independent nature of state properties represented by points 1 and 2. Changes in P, V, and T are the same for any arbitrary process leading from point 1 to 2, i.e., the values at point 2 are the same regardless of the path between the points. Isothermal (path A), isobaric (path B), and isochoric (path C) processes, represented by arbitrarily drawn lines, can be expressed in terms of partial derivatives.

Closed thermodynamic systems can be described by three state properties in which any one is a function of the other two. P, V, and T are the state variables most commonly used to define a system because P, V, and T are amenable to direct experimental measurement. If T and V are known, P is fixed and the *dependent variable P* is a state property of the *independent variables V and T*, or $P = f_1(V,T)$. Alternately, $T = f_2(V,P)$ or $V = f_3(P,T)$.

1.8 EQUATION OF STATE AND THE CONDITION OF EXACTNESS

An *equation of state* is used to quantify a state property of a specific substance in terms of other state properties. Equations of state are primarily empirically derived and can be used with confidence only within the range of measured parameters and limits of experimental error for which they were determined.

The equation of state for an *ideal gas* relates *P*,*V*, and *T* by the relationship

$$PV' = nRT \qquad [1\text{-}1]$$

where n is the number of moles. This equation is applicable at lower pressures, usually below about 5 atm, and at temperatures dependent upon the particular gas. For condensed states, three useful equations of state are the *volume thermal expansion coefficient*, α, defined by

$$\alpha = \frac{1}{V}\left(\frac{\partial V}{\partial T}\right)_P \qquad [1\text{-}2]$$

the *isothermal compressibility coefficient*, β, defined by

$$\beta = -\frac{1}{V}\left(\frac{\partial V}{\partial P}\right)_T \qquad [1\text{-}3]$$

and *Hooke's law*, defined by $\sigma = \varepsilon E$ where σ is stress, ε is strain, and E is *Young's modulus*. Values of α and β for selected substances are given in Appendix B, Tables B.1 and B.2.

Consider a system that undergoes an infinitesimal change of state. Since state changes are path independent, corresponding infinitesimals are path independent also, i.e., exact differentials. For example, if a state variable z is functionally expressed by

$$z = f(x_1, x_2, \ldots, x_n),$$

a theorem derived from the chain rule for partial derivatives then gives

$$dz = \left(\frac{\partial z}{\partial x_1}\right)_{x_2,\ldots,x_n} dx_1 + \left(\frac{\partial z}{\partial x_2}\right)_{x_1,x_3,\ldots,x_n} dx_2 + \ldots + \left(\frac{\partial z}{\partial x_n}\right)_{x_1,\ldots,x_{n-1}} dx_n \qquad [1\text{-}4]$$

where each partial derivative is itself a function of $x_1, \ldots, x_n$. The terms

$$\left(\frac{\partial z}{\partial x_1}\right)_{x_2,\ldots,x_n} dx_1, \ldots, \left(\frac{\partial z}{\partial x_n}\right)_{x_1,\ldots,x_{n-1}} dx_n$$

are called the *partial differentials* of z with respect to $x_1, x_2, \ldots, x_n$ respectively. The sum of the partial differentials, denoted as dz is the *total differential*. As an example, consider the change in internal energy of an ideal gas as a function of temperature and volume. Beginning with $U = f(V,T)$, the total differential is

$$dU = \left(\frac{\partial U}{\partial T}\right)_V dT + \left(\frac{\partial U}{\partial V}\right)_T dV.$$

For the specific case of an ideal gas, $(\partial U/\partial V)_T = 0$; hence $U = f(T)$ and only the first term is finite. For materials behavior in general, such a simplification would be incorrect. A thermodynamic property, z, that can be expressed in the form of [1-4], is defined in differential calculus as an exact differential.

An important property of exact differentials can now be derived by considering the case where $z = f(x,y)$. The result can be extended to functions of any number of independent variables. If a function f is defined by the relation $z = f(x,y)$,

$$dz = \left(\frac{\partial z}{\partial x}\right)_Y dx + \left(\frac{\partial z}{\partial y}\right)_X dy.$$

Now let $M = M(x,y) = \left(\frac{\partial z}{\partial x}\right)_Y$ and $N = N(x,y) = \left(\frac{\partial z}{\partial y}\right)_X$, then

$$dz = M(x,y)dx + N(x,y)dy.$$

Taking the partial derivative of M and N with respect to y and x respectively,

$$\left(\frac{\partial M(x,y)}{\partial y}\right)_X = \left(\frac{\partial}{\partial y}\left(\frac{\partial z}{\partial x}\right)_Y\right)_X$$

and

$$\left(\frac{\partial N(x,y)}{\partial x}\right)_Y = \left(\frac{\partial}{\partial x}\left(\frac{\partial z}{\partial y}\right)_X\right)_Y$$

The two derivatives on the right side of these last two equations are equal because the order of partial differentiation may be reversed according to a theorem from partial differential calculus (Protter and Morrey, 1970, p. 736). It follows that

$$\left(\frac{\partial M}{\partial y}\right)_X = \left(\frac{\partial N}{\partial x}\right)_Y \qquad [1\text{-}5]$$

[1-5] is known as the *condition for an exact differential* and is a necessary characteristic of a thermodynamic property. By contrast, an infinitesimal that is not the differential of a function is called an inexact differential and cannot be expressed in the form of [1-5].

1.9 PATH DEPENDENT PROCESSES: WORK AND HEAT

Unlike thermodynamic properties, *path dependent* variables are characterized by the fact that the change in the variable is contingent upon the specific path taken between states. Such variables can be thought of as process variables because numerical values are determined for the process. Path dependent variables are designated by attaching the prefix "non" to the path independent names, e.g., nonstate function or nonthermodynamic property. Work *(W)* and heat *(Q)* are two such examples. Neither can be expressed as an *exact differential*. For this reason, infinitesimals of W and Q are indicated as *inexact differentials* by using the symbol δ. The inexact differentials* of W and Q are thus written as δW and δQ respectively. It is important to note that it is improper to use such phrases as the "work contained in the body" or the "heat contained in the body." Work and heat are path dependent energy transfer processes and cannot be represented as exact differentials or point functions.

In the applications discussed in this book, only mechanical work or work against pressure will be considered. Other forms of work such as electrical, interfacial and gravitational will not be considered.

Work: Quasi-Static Tension or Compression of a Bar

When a bar, subjected to uniaxial elastic tension or compression, changes length from l to $l + dl$ where l is the original length of the bar, an infinitesimal amount of the total work performed by the axial load F applied to the bar is given by

$$\delta W' = -F\,dl \qquad [1\text{-}6]$$

In integrated form,

$$W' = -\int_{l_1}^{l_2} F\,dl$$

where F is positive for tensile loading and negative for compressive loading as shown in Figure 1.3. The minus sign is inserted here for consistency with the conventions used in this book that *work done on a system is negative*, while *work done by a system is positive*.

Since the average uniaxial *engineering stress* on a bar of *original* cross-sectional area A is $\sigma = F/A$ and the infinitesimal strain associated with dl is $d\varepsilon = dl/l$, substituting into [1-6] gives

$$\delta W' = -\sigma A l \, d\varepsilon$$

Since $Al = V'$,

$$\delta W' = -\sigma V' \, d\varepsilon \qquad [1\text{-}7]$$

---

* An alternate symbol appearing in the literature for an inexact differential is $đ$. In this book, a δ will be used to indicate an inexact differential.

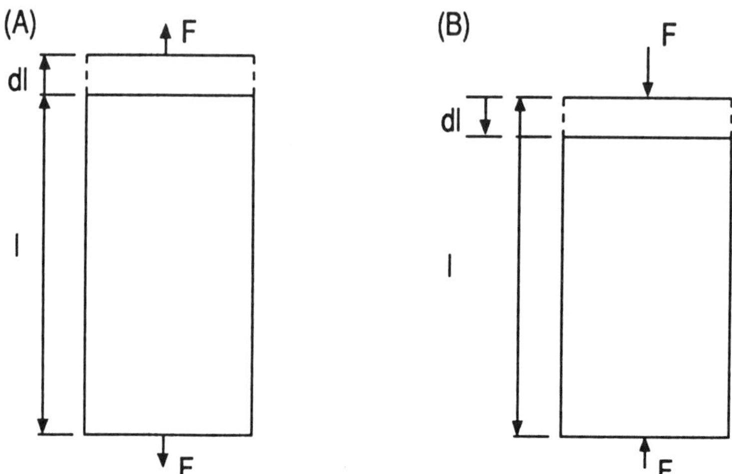

Figure 1.3 Uniaxial elastic deformation of a bar, originally of length l. (a) Tensile load F increases the length of the bar hence $W' < 0$. (b) Compressive load F decreases the length of the bar hence $W' < 0$. Negative W' agrees with the minus sign convention for the work done *on* a system.

The reader can verify the dimensional accuracy of [1-7] by substituting units for σ, V', and $d\varepsilon$.

Assuming that the change in V' during elastic deformation is small enough to be neglected, the work done per unit volume of the bar for a finite change in strain from ε_1 to ε_2 is found by integrating [1-7],

$$W = -\int_{\varepsilon_1}^{\varepsilon_2} \sigma \, d\varepsilon \quad \text{(work per unit volume)} \qquad [1\text{-}8]$$

In order to evaluate [1-8], the functional relationship between σ and ε must be known. This relationship is not readily determined if any point in the bar undergoes accelerated motion during tensile or compressive loading because both σ and ε would then be functions of time. If the external load or driving force F is slowly increased, a quasi-static process is approximated, and Hooke's law, $\sigma = \varepsilon E$, can be substituted into equation [1-8] to give

$$W = -E \int_{\varepsilon_1}^{\varepsilon_2} \varepsilon \, d\varepsilon \qquad [1\text{-}9]$$

where E, Young's Modulus, for the material comprising the bar is available for many materials. Equations [1-8] and [1-9] are not valid above the yield point of the material since irreversible deformation will occur.

Figure 1.4 illustrates experimentally derived first cycle stress-strain loops for a magnesium alloy (Dowmetal A-T4) subjected to uniaxial tensile loading. The work done per unit volume of the bar stressed from state 1 to state 2 on the original first cycle curve is the shaded area under this curve between points 1 and 2.

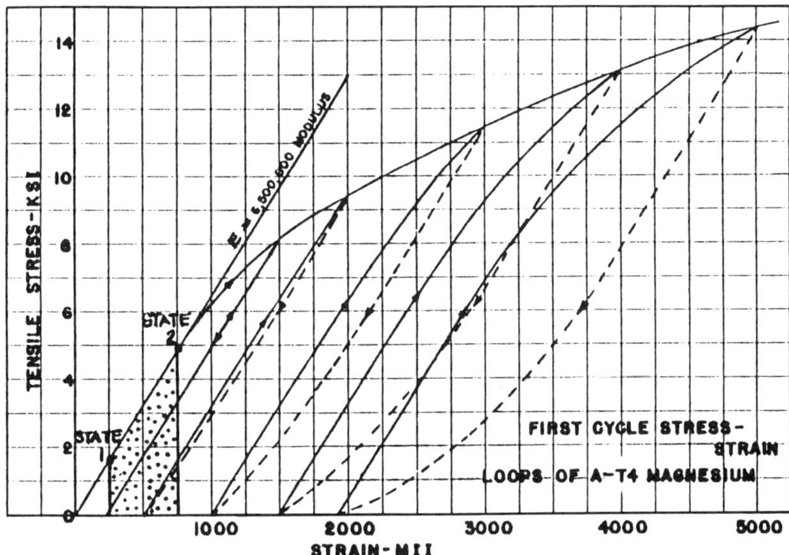

Figure 1.4 Experimentally derived uniaxial tensile first cycle stress-strain loops for Dowmetal A-T4 magnesium alloy. The work done per unit volume of the alloy during the change from state 1 to state 2 on the first loop is the shaded area shown. Equation [1-8] applies only to partial loop segments parallel to the 6.5×10^6 psi modulus slope since otherwise, irreversible permanent deformation occurs. MII = microinches per inch (From M.H. Polzin, 1951, Fig. 9 with modifications. Reprinted by permission of the Society for Experimental Mechanics, Inc., Bethel, CT.)

Example Problem 1-1

Prior to loading, the original circular cross-sectional area of a 5 m long brass bar measures 76.2 mm in diameter. After the bar is loaded in compression to 120 kip, the compressive stress is increased isothermally and quasi-statically to an engineering stress of 45 ksi.

(a) Compute the total work done as a result of increasing the stress to 45 ksi from the initial load of 120 kip. $E_{Brass} = 15 \times 10^6$ psi. State assumptions made in the calculation.

(b) What is the significance of the sign associated with the the work performed?

Solution

(a) Substituting $\varepsilon E = \sigma$ into [1-7],

$$W' = -EV' \int_{\varepsilon_1}^{\varepsilon_2} \varepsilon \, d\varepsilon = -\frac{EV'}{2}\left(\varepsilon_2^2 - \varepsilon_1^2\right). \text{ Substituting } \varepsilon = \frac{\sigma}{E},$$

$$W' = -\frac{V'}{2E}\left(\sigma_2^2 - \sigma_1^2\right).$$

Since $V' = \pi r^2 l$ for a cylinder of radius r and length l,

$$V' = 3.14 \times \left(\frac{76.2}{2 \times 25.4}\right)^2 \times 5 \times 3.281 \times 12 \text{ in}^3 = 1391 \text{ in}^3.$$

Substituting $E = 15 \times 10^6 \text{ lb}_f/\text{in}^2$, $\sigma_2 = 45 \times 10^3$ psi, and

$$\sigma_1 = \frac{120,000}{3.14} \times \left(\frac{2 \times 25.4}{76.2}\right)^2 \text{ psi} = 16,985 \text{ psi},$$

$$W' = \frac{-1391\left(45^2 - 16.985^2\right) \times 10^6}{2\left(15 \times 10^6\right) \times 12} \text{ ft} \cdot \text{lb}_f = \underline{-6710 \text{ ft} \cdot \text{lb}_f}.$$

Assumption: $V' \neq f(\varepsilon)$.

(b) $W' = -$, hence work was done *on* the system (brass bar).

Work: Quasi-Static Pressure-Volume Expansion or Compression

Figure 1.5 illustrates a *PVT* system* comprised of a gas inside a piston-cylinder arrangement. Suppose the cross-sectional area of the cylinder is A. In addition, suppose the gas in the cylinder is compressed by pushing the piston further into the cylinder and expanded when the gas pushes the piston in the opposite direction. If P is the pressure at the system-piston interface, the net force at this interface is $F = PA$. When the piston moves an infinitesimal distance dx, the work performed is equal to

$$\delta W = F \cdot dx/n \text{ or } \delta W = PA \cdot dx/n$$

---

* A *PVT* system is a *closed system* described by the macroscopic coordinates P, V, and T.

where n is the number of moles of the gas. Since $A\,dx/n = dV$,

$$\delta W = PdV \qquad\qquad [1\text{-}10]$$

Integrating [1-10],

$$W = \int_{V_1}^{V_2} PdV \qquad\qquad [1\text{-}11]$$

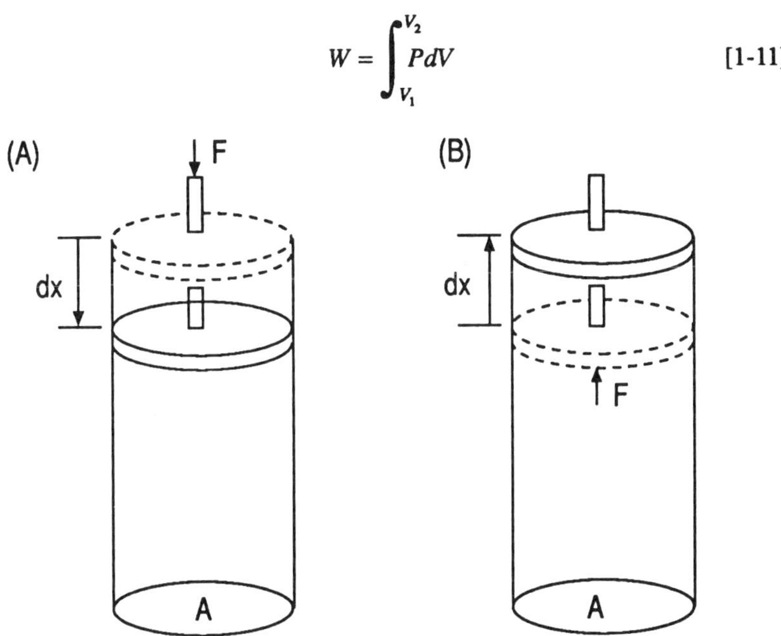

Figure 1.5 Compression of a gas (a) and expansion of a gas (b) during the motion of a piston in a frictionless cylinder of cross-sectional area A. The force, F, displaces the piston by dx. The work performed is given by [1-11].

If the piston moves with accelerated motion, both P and V are functions of time, and the computation becomes a problem in dynamics. However, if the piston moves slowly in either direction, the system approaches quasi-static expansion or compression as given by equation [1-11]. If T is constant or can be expressed as a function of V during these reversible processes, P becomes a function of V only and the calculations for work become mathematical problems involving integrals of the form

$$W = \int_{V_1}^{V_2} P(V)dV \quad (\text{T constant}) \quad \text{or} \quad W = \int_{V_1}^{V_2} P(T(V))dV \quad (\text{T} = \text{T(V)}).$$

Example Problem 1-2

Six moles of an ideal gas at 100°C (373.16 K) undergo isothermal reversible expansion against a constant external pressure of 3.5 atm in a piston-cylinder apparatus. The volume of the gas is increased by a factor of 450%.
(a) Compute the work performed as a result of the expansion.
(b) Is the + sign for work consistent with convention?

Solution
(a) The initial total volume $V_1{}'$ of the gas is found from [1-1], the equation of state for an ideal gas: $PV' = nRT$. Conversion factors and the gas constant R used in this problem are found in the Table of Physical Constants and Conversion Factors preceding Chapter 1.

$$V_1{}' = \frac{6 \times 8.3144 \times 373.16}{3.5 \times 101,325} \ \text{m}^3 = 52.492 \times 10^{-3}\ \text{m}^3.$$

The final volume V_2 is
$$V_2{}' = 4.50 \times V_1{}' = 4.5 \times 52.492 \times 10^{-3}\ \text{m}^3 = 236.214 \times 10^{-3}\ \text{m}^3.$$

$$W' = \int_{V_1'}^{V_2'} P\,dV' = P(V_2{}' - V_1{}')$$

$$= 3.5 \times 101,325(236.214 - 52.492) \times 10^{-3}$$

$$= \underline{+65,155\ \text{J.}}$$

(b) The + sign associated with W′ is consistent with convention since the gas system did work on the piston during expansion.

Work: Path Dependence

The path dependent nature of work is demonstrated in Example Problem 1-3 by considering the frictionless cylinder fitted with a movable piston illustrated in Figure 1.5. The initial state (1) and final state (4) of the gas are represented by the points (P_1, V_1, T_1) and (P_4, V_4, T_4) in Figure 1.6 respectively.

Example Problem 1-3

The purpose of this problem is two-fold. First, it illustrates the fact that work output is maximum for a reversible process. Secondly, it illustrates the fact that work output is dependent on path. In part (a), the path is chosen to be irreversible due to large and abrupt changes in P and V. In part (b), the path is chosen to be reversible or quasi-static by direct insertion of the equation of state (in this case the ideal gas law) into [1-11]. Note that the latter is closely (but not exactly) equivalent to placing a pile of sand grains on top of the piston and removing one grain at a time between state 1 and state 4. The following data applies:

    n = 1 kg mol of ideal gas, T = 25°C (constant temperature process).
    Initial pressure: P_1 = 100 atm   (State 1)
    Process I: Release to 50 atm   (State 2)
    Process II: Release to 20 atm   (State 3)
    Process III: Release to 10 atm (State 4)

(a) Calculate the work done during the irreversible expansion from state 1 to state 4.

(b) Calculate the work done during the reversible expansion from state 1 to state 4. Note that the nearly analogous removal of 1 grain of sand reduces the pressure and corresponding volume only very slightly "one grain at a time."

Solution

(a) The data are plotted on Figure 1.6.

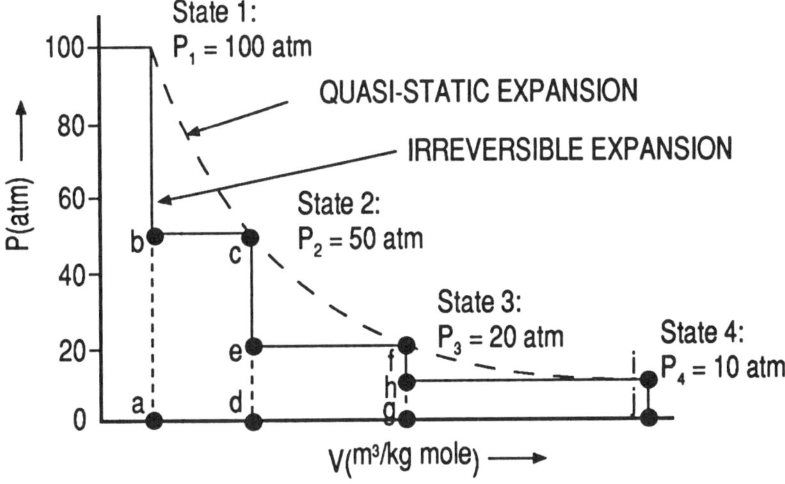

Figure 1.6 Reversible and irreversible expansion of an ideal gas.

$$V_1 = \frac{RT_1}{P_1} = \frac{8.3144 \ \text{N} \cdot \text{m} \times 298 \ \text{K} \times 1 \ \text{atm}}{\text{kg mol} \cdot \text{K} \ \times 100 \ \text{atm} \times 1.01325 \times 10^5 \ \text{N/m}^2}$$

$$= 2.45 \times 10^{-4} \ \text{m}^3/(\text{kg mol}).$$

$V_2 = V_1 P_1/P_2 = 2.45 \times 10^{-4} \times 2 = 4.9 \times 10^{-4} \ \text{m}^3/(\text{kg mol}).$
$V_3 = V_1 P_1/P_3 = 2.45 \times 10^{-4} \times 5 = 12.25 \times 10^{-4} \ \text{m}^3/(\text{kg mol}).$
$V_4 = V_1 P_1/P_4 = 2.45 \times 10^{-4} \times 10 = 24.5 \times 10^{-4} \ \text{m}^3/(\text{kg mol}).$

$$W = \int_{V_1}^{V_4} PdV = \text{area abcd} + \text{area defg} + \text{area ghij}.$$

$$= [50 \times 10^{-4}(4.9 - 2.45) + 20 \times 10^{-4}(12.25 - 4.9) + 10 \times 10^{-4} \times (24.5 - 12.25)] = 0.0392 \text{ atm·m}^3/\text{(kg mol)}$$

$$= 0.0392 \times \frac{\text{atm} \cdot \text{m}^3}{\text{kg mol}} \times \frac{1.01325 \times 10^5 \text{ N}}{\text{atm} \cdot \text{m}^2} \times \frac{1 \text{ kN} \cdot \text{m}}{1000 \text{ N} \cdot \text{m}}$$

$$= 3.972 \text{ kJ/(kg mol)}.$$

$$\text{(b)} \quad W = \int_{V_1}^{V_4} PdV = RT \int_{V_1}^{V_4 = 10V_1} dV/V = 8.3144 \times 10^{-3} \times 298 \ln(10)$$

$$= 5.705 \text{ kJ/(kg mol)}.$$

The preceeding results show that significantly more work would be accomplished if the expansion were carried out reversibly. Of course, such an expansion would be impossible because of frictional losses and time constraints.

Heat

Beginning in 1840, James Joule performed a number of experiments to determine the amount of work necessary to produce the same increase in the temperature of a system as a given amount of heat. This amount of work became known as the *mechanical equivalent of heat* and in order to determine its value, a unit of heat energy called the *calorie* was defined as the amount of heat required to raise the temperature of 1 gram of water from 14.5 to 15.5°C at 1 atm pressure. By performing mechanical as well as electrical work in adiabatically contained water, Joule determined the mechanical equivalent of heat to be 4.149 J/cal. This number was found to be completely independent of the type of work performed. The currently accepted value for the mechanical equivalent of heat (based on refined experimental techniques), known as the *thermochemical calorie*, is 4.184 J/cal.

The sign convention for describing heat transfer is: heat energy that flows *into* a body (*endothermic* process) is positive, while heat energy that flows *out* of a body (*exothermic* process) is negative.

The path-dependent nature of heat energy transfer between two thermodynamic states can be confirmed experimentally in the laboratory and is illustrated by Exercise Problem [3.5]. The similarity between Example Problem 1-3 and Exercise Problem [3.5] is illustrative of the process dependent nature of work and heat respectively, as discussed in the next chapter.

1.10 DISCUSSION QUESTIONS

(1.1) Why is classical thermodynamics thought of as a macroscopic science?

(1.2) Give examples of systems and corresponding surroundings.

(1.3) Give examples of open and closed systems.

(1.4) What is the meaning of the phrase "change in state"?

(1.5) Give several examples of systems that are chemically irreversible.

(1.6) What is a path independent property? Give examples and illustrate with drawings.

(1.7) Define a cyclic process. Give an example.

(1.8) Thermodynamic properties are "extensive" or "intensive." Define each and describe the relationship between them.

(1.9) At the beginning of what turns out to be a very hot summer, you estimate the volume of water in a pond to be 840 m^3. At the end of the summer, you estimate the volume of water to be 820 m^3. After listening to the weather report every day from the time of your first to your last volume estimate, you determine that the average barometric pressure during this time was 745 mm Hg. If the decrease in water volume was entirely the result of evaporation, is there any meaning to calculating the work done during evaporation by using the volume change and pressure conditions given here?

1.11 EXERCISE PROBLEMS

[1.1] Calculate the coefficient of thermal expansion, α, at 273 K for an ideal gas.

Ans: $\alpha = 3.66 \times 10^{-3}$ K^{-1}.

[1.2] Consider the reversible expansion process represented by the straight line 1-2 in the figure below. The working substance is 1 mole of an ideal gas. $P_1 = 1$ atm, $P_2 = 2$ atm, $V_2 = 2V_1$ and $T_1 = 300$ K. Find W for the process.

Ans: $W = 3739$ J/mol.

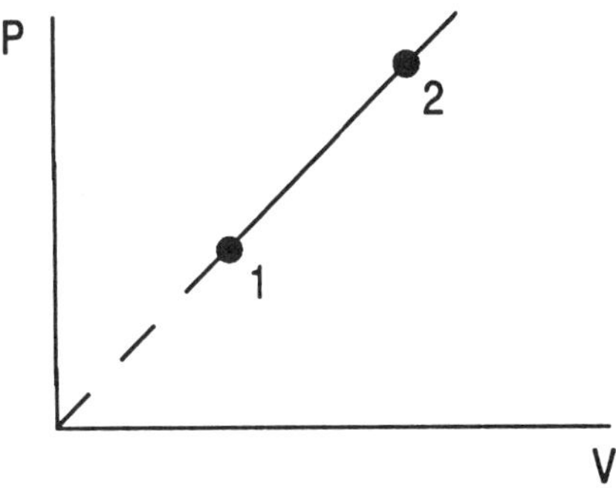

[1.3] The generalized expression

$$C_p - C_v = T\left(\frac{\partial V}{\partial T}\right)_P \left(\frac{\partial P}{\partial T}\right)_V$$

can be used to calculate $C_p - C_v$ for any substance. Show that $C_p - C_v = R$ for an ideal gas.

[1.4] An ideal gas undergoes an isobaric change from an initial State 1:$(P_1,V_1{}',T_1)$ to a final State 2:$(P_1,V_2{}',T_2)$. Prove $V_2{}' - V_1{}' = (nR/P) \times (T_2 - T_1)$.

[1.5] Show that the total work performed during the reversible isothermal expansion or compression of an ideal gas from a volume $V_1{}'$ to $V_2{}'$ is given by the equivalent expressions $W' = nRT \ln(V_2{}'/V_1{}')$ and $W' = nRT \ln(P_1/P_2)$.

[1.6] The reading on a pressure gauge relative to vacuum is added to atmospheric pressure in order to determine absolute pressure, i.e., *absolute pressure = gauge pressure + atmospheric pressure.*

(a) Calculate the work done *by* 10 moles of an ideal gas expanding reversibly from 58.784 to 14.696 psig (lb_f/in^2 gauge) in a piston-cylinder arrangement at a constant temperature of 125°F. Perform the calculation *without* using numerical gas volumes at absolute pressure. Express the answer in joules.

   Ans: $W' \approx +24{,}750$ J.

(b) Calculate the volume occupied by 10 moles of an ideal gas at 14.696 and 58.784 psig at a temperature of 125°F.

   Ans: $V'_{14.696} = 0.133$ m^3, $V'_{58.784} = 0.053$ m^3.

(c) Using the gas volumes obtained in (b), calculate the work done *by* 10 moles of an ideal gas expanding reversibly at a constant temperature of 125°F. Express the answer in joules. How does the answer compare to the one in (a)? Why are they the same or different?

   Ans: $W' \approx +24{,}750$ J, the same as (a) because the path between identical initial and final states is the same.

[1.7] Calculate the work done when the hydrostatic pressure on a cube of Cu measuring 2 cm on an edge is increased reversibly and isothermally at 0°C from 1 to 100 atm. Assume a negligible change in V over the pressure interval.

   Ans: $W \approx -0.027$ J/mol.

[1.8] Calculate the work done on the surroundings when one mole of liquid potassium expands to vapor reversibly and isothermally at 1000 K and 0.753 bar in a piston-cylinder arrangement.

   Ans: $W = 7627$ kJ/kg mol.

[1.9] Calculate the work done (internal compression) when a 350 lb mass of low carbon steel transforms from γ (austenite) to α (ferrite) during an air quench. Given that $V'_\gamma = 0.0486$ nm^3 and $2V'_\alpha = 0.0493$ nm^3, the basis for these volumes is 4 atoms of iron (one unit cell of γ or two unit cells of α [Van Vlack, 1985, p. 76]).

Ans: $W' = 30$ J.

[1.10] Calculate the isobaric temperature change necessary to produce a molar volume change of 0.02 cm^3/mol in a pyrope garnet, $Mg_3Al_2Si_3O_{12}$, crystal. Assume $\alpha V_{1\,bar,\,298\,K}$ is constant over the temperature interval.

Ans: $\Delta T = 6.71$ K.

[1.11] Calculate the molar volume of the pyroxenoid wollastonite, $CaSiO_3$, at a pressure of 200 bar and temperature of 298 K. Assume $\beta V_{1\,bar,\,298\,K}$ is constant over the pressure interval. The initial pressure is 1 bar.

Ans: $V = 39.92$ cm^3/mol.

[1.12] Express $(\partial P/\partial T)_V$ in terms of the volume thermal expansion (α) and isothermal compressibility (β) coefficients of a mineral. Assume $\Delta\alpha$ and $\Delta\beta$ are negligible for small changes in P and T.

Ans: $(\partial P/\partial T)_V = \alpha/\beta$.

[1.13] Calculate the pressure on a crystal of spinel, $MgAl_2O_4$, heated isochorically from 273 to 308 K. The pressure at 273 K is 1 bar. Assume negligible changes in α and β over the P-T interval. The volume remains constant because of external constraint.

Ans: $P_{308\,K} = 543$ bar.

[1.14] A beam in a truss bridge is subject to uniaxial tension and compression (σ) in such a manner that the beam undergoes infinitesimal changes from one *thermodynamic equilibrium* state (T_1,σ_1) to another (T_2,σ_2). The total differential of the dependent variable ε involves two parameters known from experiments to be nearly constant for small temperature changes. These parameters are the coefficient of linear thermal expansion α and Young's Modulus E. Mathematical definitions and additional experimental properties are as follows:

$$\alpha = \left(\frac{\partial\varepsilon}{\partial T}\right)_\sigma , \ \alpha \text{ is nearly independent of } \sigma$$

$$E = \left(\frac{\partial\sigma}{\partial\varepsilon}\right)_T , \ \text{for } \sigma < \text{the elastic proportional limit}$$

(a) Write an equation of state for the beam as a functional relationship.

Ans: $\varepsilon = f(T,\sigma)$.

(b) Express the exact differential of ε in terms of α and E.

Ans: $d\varepsilon = \alpha dT + (1/E)d\sigma$.

(c) Show that if ε is constant, $(\partial\sigma/\partial T)_\varepsilon = -\alpha E$.

[1.15] Beginning with the expression $P = f(V,T)$

(a) Express the total differential of P in terms of partial derivatives and name the partial differentials.

$$Ans: dP = \left(\frac{\partial P}{\partial V}\right)_T dV + \left(\frac{\partial P}{\partial T}\right)_V dT,$$

$$\text{partial differentials: } \left(\frac{\partial P}{\partial V}\right)_T dV, \ \left(\frac{\partial P}{\partial T}\right)_V dT.$$

(b) Use the condition of exactness to demonstrate that dP is an exact differential.

$$Ans: \frac{\partial^2 P}{\partial T \partial V} = \frac{\partial^2 P}{\partial V \partial T}, \text{ thus } dP \text{ is exact.}$$

[1.16] Uniaxial tensile first cycle stress-strain loops for the magnesium alloy Dowmetal A-T4 are illustrated below. Small partial cycles beginning at the lower end of each loop are on a constant slope of modulus $E = 6.5 \times 10^6$ ksi since the first part of each loop side is a line parallel to the original modulus line. For each such partial cycle, assume total elastic return upon the removal of stress. Compute the total work performed in ft · lb$_f$ when a 10.0 in. long Dowmetal A-T4 cylinder of radius 2.50 in. undergoes the partial cycles for loops 1 and 2 shown.

(a) Loop 1: $\sigma_1 = 1$ ksi, $\sigma_2 = 5$ ksi.

    $Ans: W'_1 = -30.21$ ft · lb$_f$.

(b) Loop 2: $\varepsilon_1 = 2000$ μin/in, $\varepsilon_2 = 2500$ μin/in.

    $Ans: W'_2 = -119.7$ ft · lb$_f$.

Since W'_1 and W'_2 are negative, work was done *on* the cylinder.

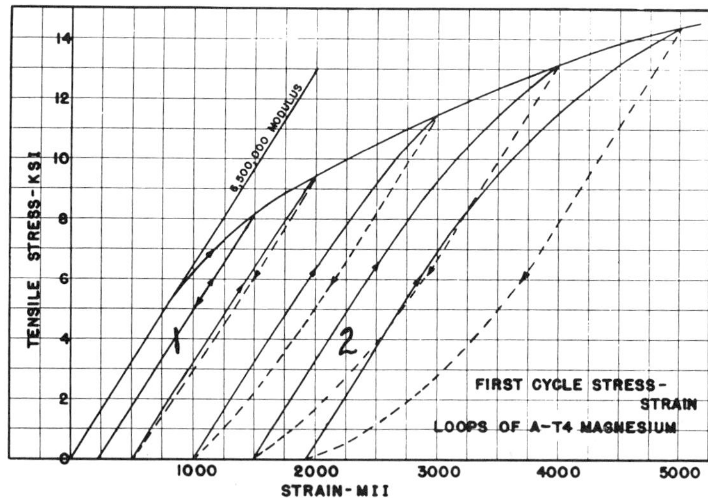

First cycle uniaxial tensile stress-strain curves for Dowmetal A-T4 magnesium alloy. Loops 1 and 2 are labeled. MII = microinches/inch. (From M.H. Polzin, 1951, Fig. 9 with modifications. Reprinted by permission of the Society for Experimental Mechanics, Inc., Bethel, CT.)

20

THE FIRST LAW OF THERMODYNAMICS AND ENTHALPY

2.1 INTERNAL ENERGY AND THE FIRST LAW OF THERMODYNAMICS

On the macroscopic scale of observation, a system possesses potential energy, E_P, by virtue of its position and kinetic energy, E_K, by virtue of its motion. On the microscopic scale, the *internal energy, U,* contained *within* a system is the sum of the potential and kinetic energies of the atoms and molecules which comprise the system. Regardless of the scale of observation, potential, kinetic, and internal energies must be measured relative to an arbitrary reference frame. Consequently, *absolute* potential, kinetic, and internal energies are undefined, and only corresponding changes have physical significance.

Let ΔE_P and ΔE_K designate changes in the macroscopic molar potential and kinetic energies of a system as a consequence of a change from state 1 to state 2. In addition, let ΔU represent the molar internal energy change corresponding to this same change in state. The total molar energy change ΔE_T of the system is then

$$\Delta E_T = \Delta E_P + \Delta E_K + \Delta U.$$

The last term in this expression, ΔU, is of fundamental importance in thermodynamics. Fortunately, the microscopic scale of observation is unnecessary to detect internal energy differences. Using only the macroscopic scale of observation, ΔU can be detected by noting differences in the macroscopic properties of a system in different states. For example, molten SiO_2 at 1000°C and 1 atm pressure has a different density, viscosity, vapor pressure, and electrical conductivity than molten SiO_2 at 1700°C and 1 atm pressure. Hence, the physical and thermodynamic properties of these two states are different.

In accordance with the conservation of energy principle, the relationship between the internal energy, work, and heat of a system must be one in which

(1) Internal energy is *increased* by work performed *on* and/or heat transferred *into* the system.

(2) Internal energy is *decreased* by work performed *by* and/or heat transferred *from* the system.

Statements (1) and (2) constitute the *First Law of Thermodynamics.* Expressed mathematically in accordance with the sign conventions for work and heat in Section 1.9, the First Law of Thermodynamics is

$$\Delta U = Q - W \qquad [2\text{-}1]$$

where Q is the molar heat absorbed or released by a system and W is the work performed per mole on or by the same system.

The First Law, written in differential form, becomes

$$dU = \delta Q - \delta W \qquad [2\text{-}2]$$

While neither Q nor W stand alone as thermodynamic properties, the difference between them is a definition of the thermodynamic property, ΔU. This difference constitutes a statement of the First Law.

Further insight into the First Law is obtained by considering two special processes involving work and heat:

(1) *Adiabatic process.* $Q = 0$; from [2-1], $\Delta U = -W_{ad}$. In such a case, W_{ad} or adiabatic work is a state function. $P, V,$ and T are variables.

(2) *Isochoric process.* $\Delta V = 0$. Since volume is constant, $W = \int P dV = 0$. From [2-1], $\Delta U = Q_v$. In such a case, Q_v is a state function. P and T are variables.

Example Problem 2-1 in Section 2.3 illustrates a general application of the First Law as well as special cases (1) and (2) for an ideal gas. The reason that an ideal gas is selected for this example is because the equation of state is simple and easy to use in illustrating a First Law application. However, caution is advised in extrapolating the results of this example to non-ideal gases or condensed states. For example, ΔU is a function of temperature only for an ideal gas and is independent of volume change. This simplification is not true in general.

2.2 ENTHALPY: A STATE FUNCTION

Substituting [1-10] into [2-2], assuming constant pressure, and integrating:

$$\int_{U_1}^{U_2} dU = Q_p - P \int_{V_1}^{V_2} dV$$

or

$$(U_2 - U_1) = Q_p - P(V_2 - V_1).$$

Upon rearranging,

$$(U_2 + PV_2) - (U_1 + PV_1) = Q_p \qquad [2\text{-}3]$$

Since $U + PV$ is an expression involving only state functions, a new thermodynamic property, *enthalpy*, is defined by

$$H = U + PV \qquad [2\text{-}4]$$

Inserting [2-4] into [2-3],

$$H_2 - H_1 = \Delta H = Q_p \qquad [2\text{-}5]$$

In differential form,

$$dH = \delta Q_p \qquad [2\text{-}6]$$

2.3 HEAT CAPACITY

Substituting [1-10] into [2-2],

$$dU = \delta Q - PdV \qquad [2\text{-}7]$$

Expressing the partial derivative form of [2-7] with respect to temperature at constant volume,

$$\left(\frac{\partial U}{\partial T}\right)_V = \left(\frac{\delta Q}{\partial T}\right)_V = C_v \qquad [2\text{-}8]$$

C_v, molar heat capacity at constant volume, is expressed in differential form by

$$dU = C_v dT \qquad [2\text{-}9]$$

or in integral form by

$$\Delta U = \int_{T_1}^{T_2} C_v dT \qquad [2\text{-}10]$$

where C_v is an intensive property (e.g., J/(mol·K)). Also,

$$dU' = C_v' dT$$

where C_v', total heat capacity at constant volume, is an extensive property (e.g., J/K).

Expressing the partial derivative form of [2-6] with respect to temperature at constant pressure,

$$\left(\frac{\partial H}{\partial T}\right)_P = \left(\frac{\delta Q_p}{\partial T}\right)_P = C_p \qquad [2\text{-}11]$$

C_p, molar heat capacity at constant pressure, is expressed in differential form by

$$dH = C_p dT \qquad [2\text{-}12]$$

or in integral form by

$$\Delta H = \int_{T_1}^{T_2} C_p dT \qquad [2\text{-}13]$$

where C_p is an intensive property (e.g., J/(mol·K)). Also,

$$dH' = C_p' dT$$

where C_p', total heat capacity at constant pressure, is an extensive property

(J/K). It is left to the reader to develop expressions for heat capacity on a mass basis* (i.e., specific heat capacities c_v and c_p at constant volume and pressure respectively).

Example Problem 2-1

Consider a quasi-static three step process. The working substance is 1 lb mol of diatomic ideal gas initially at 70°F and 1 atm. The gas is heated at constant volume to 400°F, expanded adiabatically to the initial temperature of 70°F, and finally compressed isothermally to the initial pressure of 1 atm. Confirm the data table illustrated at the end of the solution below.

Solution

Construct a "closed loop" or graphical scheme of the given and required data. The process and given data are illustrated in Figure 2.1. The method of solution involves three general steps: (a) compute all P-V-T data possible from the ideal gas equation of state: $PV = RT$ ($n = 1$), (b) compute process paths I, II, and III; (c) confirm the data table. Note that the cyclic sum for each process serves as a data check.

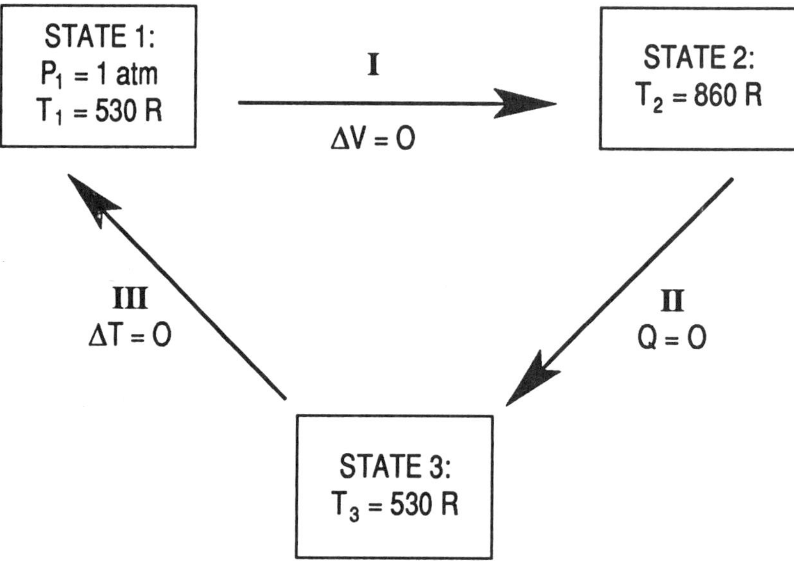

Figure 2.1 Quasi-static cyclic process corresponding to the data given in Example Problem 2-1. The data are labeled on the diagram. I: isochoric heating, II: adiabatic expansion; III: isothermal compression.

---

* The kinetic theory of gases predicts heat capacity values for ideal gases. For example, for a monatomic gas: $C_p = (5/2)R$, $C_v = (3/2)R$, and for a diatomic gas: $C_p = (7/2)R$, $C_v = (5/2)R$. When the specific gas is not specified, these values should be used.

(a) Compute all *P-V-T* data from the ideal gas equation.
State 1: $P_1V_1 = RT_1$.

$$V_1 = \frac{RT_1}{P_1} = \frac{0.73 \times 530}{1}\,\text{ft}^3/(\text{lb}\cdot\text{mol})$$

$$= 386.9\ \text{ft}^3/(\text{lb}\cdot\text{mol})$$

State 2: $V_2 = V_1 = 386.9\ \text{ft}^3/(\text{lb}\cdot\text{mol})$

$$P_2V_2 = RT_2$$

$$P_1V_1 = RT_1$$

$$P_2 = \frac{T_2}{T_1} \times P_1 = \frac{860}{530} \times 1\ \text{atm}$$

$$= 1.623\ \text{atm}.$$

State 3: Since *P*, *V*, and *T* vary when $Q = 0$, relationships specific to an adiabatic process are applicable. These relationships are derived as follows. Beginning with $U = f(V,T)$ and using the chain rule,

$$dU = \left(\frac{\partial U}{\partial T}\right)_V dT + \left(\frac{\partial U}{\partial V}\right)_T dV.$$

Since $U = f(T)$ only for an ideal gas, $(\partial U/\partial V)_T\, dV = 0$. Substituting [2-8] into the chain rule expression above and combining with [1-10] and [2-2],

$$dU = C_v dT = -P dV.$$

Differentiating the ideal gas equation of state $PV = RT$,

$$PdV + VdP = RdT,$$

and substituting into the above expression for *dU*,

$$dU = C_v dT = VdP - RdT.$$

Separating variables and substituting for *V* from [1-1] for $n = 1$,

$$(C_v + R)\frac{dT}{T} = R\frac{dP}{P}.$$

Integrating,

$$\left(\frac{C_V + R}{R}\right)\int_{T_2}^{T_3} \frac{dT}{T} = \int_{P_2}^{P_3} \frac{dP}{P}$$

$$\left(\frac{T_3}{T_2}\right)^{\frac{(C_v + R)}{R}} = \frac{P_3}{P_2}$$

or
$$P_3 = P_2\left(\frac{T_3}{T_2}\right)^{\gamma/(\gamma-1)}$$

where $\gamma = C_p/C_v = 7/5$ ($C_p = (7/2)R$ for a diatomic ideal gas and $C_p - C_v = R$). Hence,

$$P_3 = 1.623\left(\frac{530}{860}\right)^{\frac{7/5}{[(7/5)-1]}} = 1.623\left(\frac{530}{860}\right)^{7/2} = 0.298 \text{ atm.}$$

Using $P_3V_3 = RT_3$,

$$V_3 = \frac{RT_3}{P_3} = \frac{0.73 \times 530}{0.298} = 1298 \text{ ft}^3/(\text{lb} \cdot \text{mol}).$$

(b) Compute remaining data for process paths I, II, and III.

Path I: $\Delta U = Q - W = Q - 0$ (isochoric)

$$= C_v \int_{T_1}^{T_2} dT = (5/2)R \int_{530}^{860} dT = (5/2)(1.987)(330)$$

$$= 1639 \text{ BTU/(lb mol).}$$

Path II: $\Delta U = Q - W = 0$ (adiabatic) $- W$

$$= C_v \int_{T_2}^{T_3} dT = (5/2)R \int_{860}^{530} dT = (5/2)(1.987)(-330)$$

$$= -1639 \text{ BTU/(lb} \cdot \text{mol) and } W = 1639 \text{ BTU/(lb} \cdot \text{mol).}$$

Path III: $\Delta U = Q - W = 0$ (isothermal)

$$Q = W = \int_{V_3}^{V_1} P\,dV = RT \int_{1298}^{386.9} \frac{dV}{V}$$

$$= 1.987 \times 530 \times \ln\left(\frac{386.9}{1298}\right)$$

Thus $Q = W = -1275$ BTU/(lb · mol).

(c) The completed data table is illustrated below.

Path	ΔU BTU/(lb mol)	Q BTU/(lb mol)	W BTU/(lb mol)	ΔP atm	ΔT R	ΔV ft³/(lb mol)
I	1639	1639	0	+0.623	+330	0
II	−1639	0	1639	−1.325	−330	+911.1
III	0	−1275	−1275	+0.702	0	−911.1
Cycle	0	364	364	0	0	0

2.4 ENTHALPY (HEAT) OF TRANSFORMATION

The term *phase transformation* refers to either a polymorphic solid-solid transformation, a solid-liquid transformation, or a condensed phase-gas transformation. A *polymorphic (allotropic) transformation* occurs when a solid element or compound undergoes a change in crystal structure. There is no change in the chemistry of the solid, and each structural variety of the solid is called a *polymorph* of that solid.

The phase boundaries of Figure 2.2, for example, illustrate the range of pressures and temperatures at which various SiO_2 polymorphic transformations occur. At 1713°C or above, *fusion* occurs and SiO_2 melts to form liquid. Figure 2.3 illustrates the equilibrium phase diagram for pure H_2O. The range of pressures and temperatures over which fusion occurs is indicated by the inclined solid-liquid phase boundary.

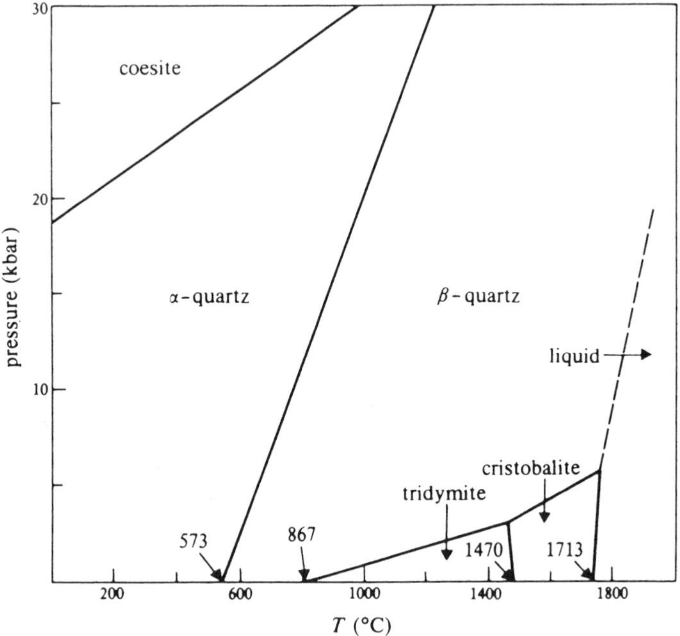

Figure 2.2 Polymorphs of SiO_2. (After F.R. Boyd and J.L. England, 1960.)

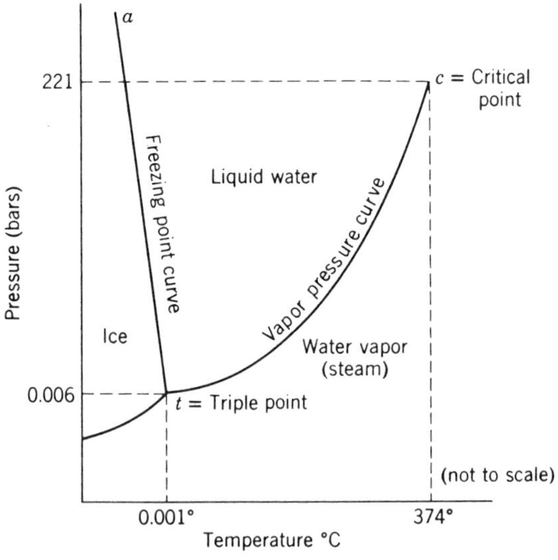

Figure 2.3 A portion of the equilibrium phase diagram of H_2O, a one component system. (From C. Klein and C.S. Hurlbut, Jr., 1985, Manual of Mineralogy, Fig. 4.37. Reprinted by permission of John Wiley and Sons, Inc., Copyright © 1985.)

Two types of condensed phase-gas transformations are shown by the concave upward curves. Above the "triple point" pressure, liquid evaporates to form water vapor. Below the "triple point" pressure, solid evaporates to form water vapor by the process of *sublimation*.

The pressures and temperatures at which a phase transformation occurs are determined experimentally. For example, for a one-component system at constant pressure, cooling curve plots of temperature versus time reveal a "flat" or isothermal hold during the time interval between the start and finish of a liquid to solid transformation, Figure 2.4. The "hold temperature" is identified as the transformation temperature provided the cooling is slow enough to prevent supercooling or non-equilibrium cooling. Heat is released ($Q_p = -\Delta H^f$ where ΔH^f is the *heat* or *enthalpy of fusion*) during the hold and the temperature cannot decrease until all liquid is solidified. Recalling [2-5], the terms *heat of transformation* and *enthalpy of transformation* are often used interchangeably. Hence for any transformation at constant pressure,

$$\Delta H^{Tr} = Q_p.$$

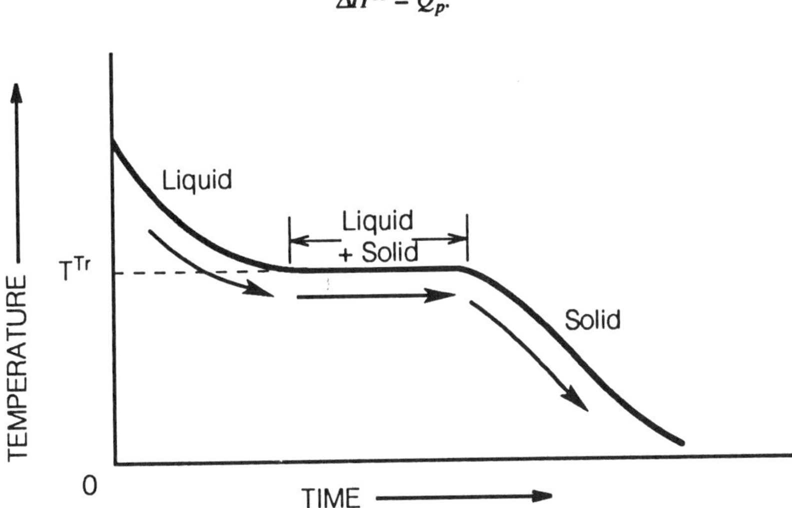

Figure 2.4 Hypothetical temperature-time equilibrium cooling curve for a one component system. At the transformation temperature, T^{Tr}, a liquid to solid transformation occurs.

The effect of supercooling on $-\Delta H^f$ is illustrated later in this chapter. In addition, the phase rule and its application to two component phase diagrams is introduced in Chapter 7.

While Figure 2.3 is only qualitative, vapor-liquid equilibrium data are published for a variety of working substances used in power system design. For example, Appendix D, Table D.1, lists typical vapor-liquid equilibrium data for water. Based on this data, water exerts a vapor pressure of 0.25 bar ($\approx$0.25

atm) at 65°C. The heat absorbed when a phase change occurs is $\Delta h^{Tr} = h_g - h_l$ = 2618 − 272 kJ/kg = 2346 kJ/kg. A comparison of data in Appendix D, Table D.2, for potassium at ≈ 0.25 bar shows a substantially higher equilibrium temperature of ≈ 627°C but a similar $\Delta h^{Tr} \approx 2009$ kJ/kg. The engineering-materials aspects of these comparisons are significant as will be discussed later. Heats of transformation for selected substances are found in Appendix A, Table A.2.

2.5 STANDARD STATE ENTHALPY

Since energy in any form is relative, only changes in enthalpy can be computed or measured. Hence, an arbitrary reference or standard state has been universally accepted. The *standard state enthalpy* or *heat content* of an element or compound is defined for the most *stable* form of that element or compound under the chosen conditions of temperature and pressure. It is denoted by superscript 0 and is defined in Table 2.1 for solids, liquids, and gases in elemental or compound form. By convention, the heat content of an element is zero at 298 K and one atmosphere pressure. The heat of formation of a compound, denoted by $\Delta H_T^{0,f}$ in Table 2.1, is defined in the next section.

Table 2.1: Standard State Enthalpy

State of Aggregation	*Pure Elements* $P^0 = 1$ atm* Stable[†] at 298 K	*Pure Compounds* $P^0 = 1$ atm* Stable[†] at T (K)
Solid	$\Delta H_{298}^0 = 0$	$\Delta H_T^0 = \Delta H_T^{0,f}$
Liquid	$\Delta H_{298}^0 = 0$	$\Delta H_T^0 = \Delta H_T^{0,f}$
Gas (ideal)	$\Delta H_{298}^0 = 0$	$\Delta H_T^0 = \Delta H_T^{0,f}$

* *Standard state pressure* is an arbitrarily adopted unit and some tables of thermodynamic data (e.g., Robie et al., 1979 and Holland, 1990) are based on a standard pressure of 1 bar (1 bar = 1.01325 atm). Changing the reference pressure from 1 atm to 1 bar has a negligible effect on the thermodynamic properties tabulated in this book.

[†] For diatomic gases such as hydrogen, H_2, $\Delta H_{298}^0 = 0$. For monatomic hydrogen, H, $\Delta H_{298}^0 \neq 0$.

2.6 ENTHALPY (HEAT) OF FORMATION AND REACTION
Heat of Formation

Heat or enthalpy of formation, $\Delta H_T^{0,f}$, is defined as the heat associated with the formation of a *compound* from its *elements* at *any temperature* at which the compound and its component elements are *stable*. Hence, at $P^0 = 1$ atm and $T = 298$ K for which data is readily available in the literature,

$$\int_{\text{Reactants}}^{\text{Products}} dH = \Delta H_{298}^{0,f} = \Delta H_{298}^{0} \text{ (Product)} - \Sigma n \Delta H_{298}^{0} \text{ (Elements)}$$

or $\Delta H_{298}^{0,f} = \Delta H_{298}^{0}$ (Product)

where n is the number of moles of each element and $\Delta H_{298}^{0,f}$ is the *standard state heat of formation*. Values of $\Delta H_{298}^{0,f}$ for selected compounds are given in Appendix A, Table A.1. From Table 2.1, $\Sigma n \Delta H_{298}^{0}$ (Elements) = 0.

Heat of Reaction

The *heat or enthalpy of reaction, ΔH_T,* at temperature T is defined as the heat released or absorbed when reactants form products in a *balanced* chemical reaction.

Hence, at $P^0 = 1$ atm. and $T = T(\text{K})$,

$$\int_{\text{Reactants}}^{\text{Products}} dH = \Delta H_T^{0}$$

or $\qquad\qquad \Delta H_T^{0} = \Sigma n \Delta H_T^{0,f} \text{ (Products)} - \Sigma n \Delta H_T^{0,f} \text{ (Reactants)} \qquad\qquad$ [2-14]

where ΔH_T^{0} is the standard heat of reaction and n is the number of moles of each reactant or product. Note that the superscript f applies to compounds but not to elements.

As an example of the application of [2-14], consider the chemical reaction occurring in the standard state ($P^0 = 1$ atm, $T = T(\text{K})$) for reactants A and B forming products C and D:

$$n_A A + n_B B \rightarrow n_C C + n_D D.$$

The standard state heat of reaction is

$$\Delta H_T^{0} = n_C \Delta H_{T,C}^{0,f} + n_D \Delta H_{T,D}^{0,f} - n_A \Delta H_{T,A}^{0,f} - n_B \Delta H_{T,B}^{0,f}$$

where $\Delta H_{T,A}^{0,f}$, $\Delta H_{T,B}^{0,f}$, $\Delta H_{T,C}^{0,f}$, and $\Delta H_{T,D}^{0,f}$, are the standard state molar heats or enthalpies of formation of A, B, C, and D respectively, at temperature T. If A, B, C, or D are elements at 298 K, then $\Delta H_{298}^{0,f} = 0$. The *number of moles* (corresponding to n_A, n_B, n_C, and n_D) in the chemical reaction are incorporated into the summation formula. Note that ΔH_T^{0} is written on a per-mole basis corresponding to n_A moles of A, n_B moles of B, etc. If $\Delta H_T^{0} = 50$ kJ/mol, then ΔH_T^{0} is 50 kJ/n_A = (50/n_A) kJ/mol of A, 50 kJ/n_B = (50/n_B) kJ/mol of B, etc.

It is possible to compute the enthalpy change of a reaction even if all of the enthalpies of formation of the products and reactants are unknown. Suppose the reaction in question is written as a combination of two or more reactions for which enthalpy changes are known at the *same* pressure and temperature as the desired reaction. The enthalpy change of the desired reaction can then be computed by combining the enthalpies of the constituent reactions. The enthalpies are combined in a manner analogous to that in which the reactions themselves must be combined to produce the desired reaction. This important fact, known as *Hess's law*, applies to all *thermodynamic* functions. Application of Hess's law to calculation of standard state reaction enthalpy is illustrated in Example Problem 2-2.

Example Problem 2-2

Given: $T = 298$ K, $P = 1$ atm and

(1) $W(s) + O_2(g) = WO_2(s)$ $\qquad \Delta H_{298}^{0,f} = -560.7$ kJ/mol O_2

(2) $WO_2(s) + O_2(g) = W_3O_8(s)$ $\qquad \Delta H_{298}^{0} = -550.2$ kJ/mol O_2

(3) $W_3O_8(s) + (1/2)O_2(g) = 3WO_3(s)$ $\qquad \Delta H_{298}^{0} = -278.3$ kJ/mol W_3O_8

Find: $\Delta H_{298}^{0,f}, WO_3$.

Solution

Equations and enthalpies are additive.

(1) $W + O_2 = WO_2$ $\qquad \Delta H_{298}^{0,f} = -560.7$ kJ/mol O_2

(2) $WO_2 + (1/3)O_2 = (1/3)W_3O_8$ $\qquad \Delta H_{298}^{0} = -183.4$ kJ/mol WO_2

(3) $(1/3)W_3O_8 + (1/6)O_2 = WO_3$ $\qquad \Delta H_{298}^{0} = -92.7$ kJ/mol WO_3

(4) $(1) + (2) + (3) = (4)$

or $W(s) + (3/2)O_2(g) = WO_3(s)$ $\qquad \Delta H_{298,WO_3}^{0,f} = \underline{-836.8 \text{ kJ/mol W}}$.

Example Problem 2-3

The standard enthalpies of formation of several minerals at 968 K are as follows:

(a) $Al_6Si_2O_{13}$ (mullite) $\qquad \Delta H^{0,f} = 42.2$ kJ/mol

(b) Al_2O_3 (α-corundum) $\qquad \Delta H^{0,f} = 31.8$ kJ/mol

(c) SiO_2 (quartz) $\qquad \Delta H^{0,f} = -15.3$ kJ/mol

Calculate ΔH^0 for the production of mullite from α-corundum and quartz at 968 K.

Solution

The balanced reaction is

$$3Al_2O_3(\alpha\text{-corundum}) + 2SiO_2(\text{quartz}) \rightarrow Al_6Si_2O_{13}(\text{mullite}).$$

The enthalpy change of the reaction is calculated from [2-14],

$$\Delta H_{968}^0 = (1)\,\Delta H_{\text{mull}}^{0,f} - (3)\,\Delta H_{\alpha\text{-cor.}}^{0,f} - (2)\,\Delta H_{\text{qz}}^{0,f}.$$

Substituting the formational enthalpy values for the products and reactants,

$$\Delta H_{968}^0 = \{(1)(42.2) - (3)(31.8) - (2)(-15.3)\}\ \text{kJ/mol.}$$

Since $n = 1$ for mullite,

$$\Delta H_{968}^{0} = \underline{-22.6\ \text{kJ/mol mullite.}}$$

Note that ΔH_{968}^0 is not equal to $\Delta H_{\text{mull}}^{0,f}$ in this example because mullite in the above reaction is not produced from its elements.

2.7 INTRODUCTION TO THE THERMODYNAMIC LOOP (TL)

Although Hess's law is useful, it is limited in application to constant pressure and temperature reactions. A more general and useful analytical tool is direct application of Kirchhoff's law to solve a wide variety of thermodynamic problems. Kirchhoff's law is referred to in this book by the term *thermodynamic loop* (TL). The TL represents an effective way to organize and simplify the solution of more complex problems in chemical thermodynamics (personal communication, 1959, R. Schuhmann, Jr., Department of Metallurgical Engineering, Purdue University, West Lafayette, Indiana). The TL approach is described (Stracher and Johnson, 1990) by the property

$$\sum \Delta\theta\bigg|_{\text{Loop}} = 0$$

where $\Delta\theta$ is the change in value of the state function θ between any two points along the loop (see, for example, Fine and Geiger, 1979, and Hamill et al., 1966). Figure 2.5 illustrates a thermodynamic loop with a *state function path* defined by the property $\sum\Delta\theta_{TL} = 0 = \Delta\theta_{A\rightarrow B} + \Delta\theta_{B\rightarrow C} + \Delta\theta_{C\rightarrow A}$. In addition, the path taken between any two points along any segment of the loop results in the same $\Delta\theta$. For example, $\Delta\theta_{A\rightarrow B,\,\text{path 1}} = \Delta\theta_{A\rightarrow B,\,\text{path 2}}$. As an illustration of thermodynamic loop analysis, an expression will be derived for the enthalpy change of a reaction as a function of temperature in the following problem.

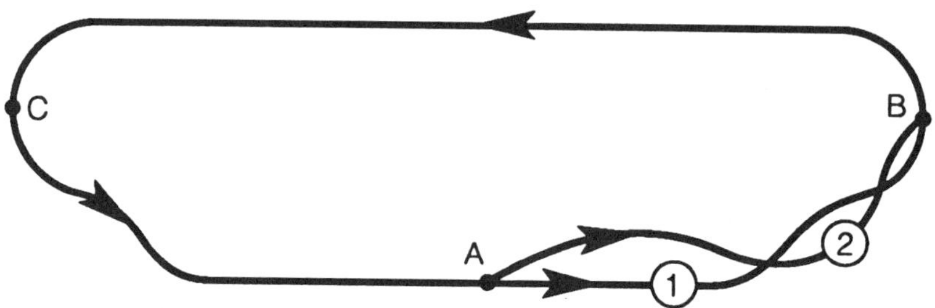

Figure 2.5 Thermodynamic loop with state function path A-A. (From G.B. Stracher and D.L. Johnson, 1990.)

Example Problem 2-4

Derive an expression for the standard enthalpy change of the general chemical reaction $n_A A + n_B B \rightarrow n_C C + n_D D$ as a function of temperature at $P^0 = 1$ atm.

Solution

An isobaric thermodynamic loop is set up as follows:

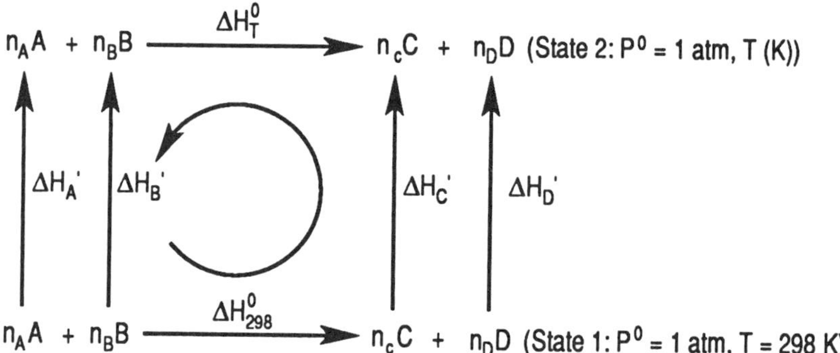

ΔH_{298}^0 and ΔH_T^0 are *reaction enthalpies* for states 1 and 2 respectively. $\Delta H_A'$, $\Delta H_B'$, $\Delta H_C'$, and $\Delta H_D'$ are the *total enthalpy* changes from state 1 to 2 for reaction components A, B, C, and D respectively. Summing around the loop in an *arbitrary counterclockwise* direction,

$$\Sigma \Delta H_{TL} = 0 = \Delta H_{298}^0 + \Delta H_C' + \Delta H_D' - \Delta H_T^0 - \Delta H_A' - \Delta H_B'.$$

Applying [2-13] to each reaction component and rearranging,

$$\Delta H_T^0 = \Delta H_{298}^0 + n_C \int_{298}^{T} C_p^C dT + n_D \int_{298}^{T} C_p^D dT - n_A \int_{298}^{T} C_p^A dT - n_B \int_{298}^{T} C_p^B dT$$

or

$$\Delta H_T^0 = \Delta H_{298}^0 + \int_{298}^{T} \Delta C_p dT \qquad [2\text{-}15]$$

In general,

$$\Delta H_T^0 = \Delta H_{T_1}^0 + \int_{T_1}^{T} \Delta C_p dT \qquad [2\text{-}16]$$

where $\Delta C_p = n_C C_p^C + n_D C_p^D - n_A C_p^A - n_B C_p^B$. For a reaction involving any number of components,

$$\Delta C_p = \sum n_i C_p^i \text{ (Products)} - \sum n_i C_p^i \text{ (Reactants)} \qquad [2\text{-}17]$$

where n_i and C_p^i are the number of moles of component i in the reaction and the molar heat capacity of component i respectively.

Equations [2-15] and [2-16] are normally found in thermodynamics textbooks for the case in which no phase changes occur. Phase changes are incorporated into the TL in Example Problem 2-5.

The solution to [2-16] now becomes one of evaluating C_p for each component in the reaction. In Appendix A, Table A.3A, C_p is expressed by the empirical equation

$$C_p = a + bT + cT^{-2} + dT^{-0.5} \qquad [2\text{-}18]$$

where a, b, c and d are constants. Examination of [2-16] and [2-18] reveals two simplifying cases:

Case I: If $a = b = c = d = 0$, $\Delta C_p = 0$, hence

$$\Delta H_T^0 = \Delta H_{T_1}^0 \qquad [2\text{-}19]$$

Case II: If $b = c = d = 0$, $\Delta C_p = a$ (constant), hence

$$\Delta H_T^0 = \Delta H_{T_1}^0 + a(T - T_1) \qquad [2\text{-}20]$$

It should be noted that for many engineering applications, the assumption $\Delta C_p \approx 0$ simplifies computations without compromising the engineering accuracy of the result. Further reference to these simplifying cases will be made in subsequent chapters.

A suggested procedure for organizing a thermodynamic loop can be remembered in terms of *The Four S's Of Thermodynamic Loop Analysis:* (1) Setup, (2) Sum, (3) Substitute, and (4) Solve. Each procedure is performed as follows:

(1) **Set Up:** Write down the reaction for which data is known (typically the standard state), balance and label it state 1. Write down the same reaction for which information is desired and label it state 2. Connect the corresponding reactants and products in each state with vertical arrows. The direction of these arrows (up or down) is arbitrary. If a reactant or product changes phase between states 1 and 2, the phase change must be incorporated into the proper loop segment.

(2) **Sum:** The direction chosen for summation is arbitrary and is indicated by a single clockwise or counterclockwise arrow drawn between the reactions for each state. Counterclockwise summation is used for all examples in this book, hence, the right hand is used with the thumb pointing up. The remaining fingers of the right hand are curled in the direction chosen for summation. Sum properties around the loop and set the sum equal to zero. Loop segments pointing in the summation direction imply a plus sign in the sum, and loop segments pointing opposite to the summation direction imply a minus sign in the sum. Rearrange the resulting equation and solve algebraically for the desired unknown.

(3) **Substitute:** Obtain numerical data from the problem statement and/or the literature and incorporate this data into the appropriate loop segments. Although data may be readily available in tables, charts, or nomographs, a more extensive search may be required for complex nonmetallic and metallic systems. Carefully examine the problem statement to insure that no important restrictions or assumptions are overlooked.

(4) **Solve:** Solve the derived equation for the unknown.

This chapter concludes with numerical examples using TL analysis.

Example Problem 2-5

Calculate the standard enthalpy of formation of TiC at 1200 K. The data necessary to solve this problem is tabulated in Tables A.1, A.2, and A.3A.

Solution

(1) **Setup.** A TL incorporating the reaction in question is structured to include the Ti(α) $\rightarrow$ Ti(β) polymorphic transformation at 1155 K as follows:

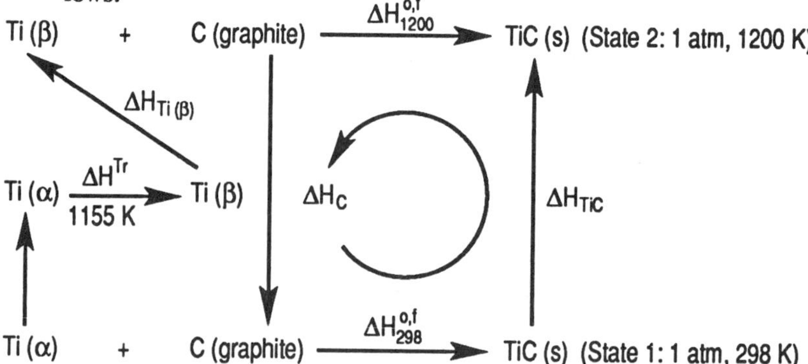

(2) Sum. Summing in the counterclockwise direction,

$$\Sigma \Delta H_{TL} = 0 = \Delta H_{298}^{0,f} + \Delta H_{TiC} - \Delta H_{1200}^{0,f} + \Delta H_C - \Delta H_{Ti(\beta)} - \Delta H^{Tr} - \Delta H_{Ti(\alpha)}.$$

Rearranging and solving for $\Delta H_{1200}^{0,f}$,

$$\Delta H_{1200}^{0,f} = \Delta H_{298}^{0,f} + \Delta H_{TiC} + \Delta H_C - \Delta H_{Ti(\beta)} - \Delta H^{Tr} - \Delta H_{Ti(\alpha)}.$$

(3) Substitute. Inserting the heat capacity and transformation enthalpy data into the above expression,

$$\Delta H_{1200}^{0,f} \text{ (J/mol)} = -183.70 \times 10^3$$

$$+ \int_{298}^{1200} (1)(49.50 + 3.35 \times 10^{-3} T - 14.98 \times 10^5 T^{-2}) dT$$

$$+ \int_{1200}^{298} (1)(17.15 + 4.27 \times 10^{-3} T - 8.79 \times 10^5 T^{-2}) dT$$

$$- \int_{1155}^{1200} (1)(28.91) dT - (1)3473 - \int_{298}^{1155} (1)(22.09 + 10.04 \times 10^{-3} T) dT$$

(4) Solve. Integration, left to the reader, gives

$$\Delta H_{1200}^{0,f} = \underline{-186.7 \text{ kJ/mol}}$$

The negative enthalpy of formation of TiC in state 2 implies that the reaction is exothermic. Since both state 1 and state 2 reactions occur at constant pressure (P^0 = 1 atm), enthalpy changes are equal to the heat evolved. For state 1, the heat evolved is $Q_1 = -183.7$ kJ/mol and for state 2 the heat evolved is $Q_2 = -186.7$ kJ/mol; hence, more heat is evolved at the higher temperature. Note that for each integral, the *lower* temperature limit of integration corresponds to the temperature of the phase at the *tail* of each arrow, while the *upper* temperature limit of integration corresponds to the temperature of the phase at the *head* of each arrow. In addition, as with Hess's law, it is necessary in thermodynamic loop analysis that *all* reactions be properly balanced.

Example Problem 2-6

Small droplets of gold are observed to *supercool* (remain liquid) 230°C below the normal freezing point at atmospheric pressure. Calculate the transformation enthalpy or heat of transformation associated with solidification in the supercooled state. The data necessary to solve this problem is tabulated in Tables A.2 and A.3A.

Solution
Since the heat of solidification is the opposite of the heat of fusion, $\Delta H(l \rightarrow s)$ $= -\Delta H(s \rightarrow l) = -\Delta H^f$. The directions of the arrows in the TL are set up to yield $\Delta H(l \rightarrow s)$ directly. Since C_p for liquid gold at 1 atm pressure is available only from the melting point to 1600 K, C_p is extrapolated to 1106 K in order to solve this problem.

(1) Set Up.

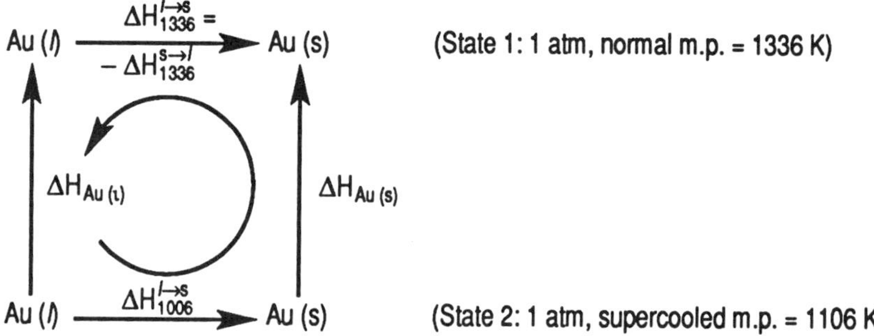

(2) Sum. Summing in the counterclockwise direction,

$$\Sigma \Delta H_{TL} = 0 = \Delta H_{1106}^{l \rightarrow s} + \Delta H_{Au(s)} - \left(-\Delta H_{1336}^{s \rightarrow l}\right) - \Delta H_{Au(l)}$$

Rearranging and solving for $\Delta H_{1106}^{l \rightarrow s}$,

$$\Delta H_{1106}^{l \rightarrow s} = -\Delta H_{Au(s)} - \Delta H_{1336}^{s \rightarrow l} + \Delta H_{Au(l)}.$$

(3) Substitute. Inserting heat capacity and transformation enthalpy data into the above expression,

$$\Delta H_{1106}^{l \rightarrow s}\ (\text{J/mol}) = -\int_{1106}^{1336} (1)(23.68 + 5.19 \times 10^{-3}T)dT - 12.76 \times 10^3$$
$$+ \int_{1106}^{1336} (1)29.29\,dT.$$

(4) Solve. Integrating,

$$\Delta H_{1106}^{l \rightarrow s} = -23.68(1336 - 1106) - (5.19 \times 10^{-3}/2)$$
$$\times (1336^2 - 1106^2) - 12.76 \times 10^3 + 29.29(1336 - 1106)$$

which reduces to

$$\Delta H_{1106}^{l \to s} = -\underline{12.93 \text{ kJ/mol}}.$$

Since the phase changes occur at constant pressure (1 atm), the heat of transformation for state 2 is

$$Q^{\text{Tr}} = \Delta H_{1106}^{l \to s} = -\underline{12.93 \text{ kJ/mol}}.$$

The liquid to solid phase changes occurring in states 1 and 2 result in negative transformation enthalpies or heats of transformation. This change implies that heat energy is released to the surroundings as liquid gold solidifies. Note that the difference in $\Delta H_{Au}^{l \to s}$ between 1106 K and 1336 K is only 170 J/mol or 1.3%.

2.8 DISCUSSION QUESTIONS

(2.1) Can the change in a state function be equal to the difference of two nonstate functions? Give an example.

(2.2) The internal energy change of a system is related to the performance of adiabatic work as follows:
(1) $\Delta U = -W_{\text{Ad.}} > 0$
(2) $\Delta U = -W_{\text{Ad.}} < 0$.
In which case is work done *by* the system? Explain.

(2.3) Derive the formulas listed below for a *PVT* system.
(a) Constant volume process, *PV* work only:

$$\Delta U' = Q_v' = \int_{T_1}^{T_2} mc_v dT .$$

(b) Constant pressure process, PV work only:

$$\Delta H' = Q_p' = \int_{T_1}^{T_2} mc_p dT .$$

(2.4) Many compounds have two or more polymorphs. For example, SiO_2 has 6 polymorphs shown in Figure 2.2, while the two polymorphs of $CaCO_3$ are the high pressure variety aragonite and the low pressure variety calcite. Give examples of polymorphic phase transformations using Figure 2.2.

(2.5) Two phase transformations occur at a temperature T and pressure P. The phases involved are A and A', and the transformations and enthalpy changes are:
(1) $A \xrightarrow[(T,P)]{} A'$ $\Delta H_{(1)} = Q_{(1)}$
(2) $A' \xrightarrow[(T,P)]{} A$ $\Delta H_{(2)} = Q_{(2)}$
How is $\Delta H_{(1)}$ related to $\Delta H_{(2)}$? How is $Q_{(1)}$ related to $Q_{(2)}$?

(2.6) From the definition of standard state enthalpy of a diatomic gas at 298 K,

(a) Is $\Delta H^0_{N(g)} = 0$?

(b) Is $\Delta H^0_{N_2(g)} = 0$?

Explain both answers.

(2.7) For which of the reactions below is $\Delta H^0 = \Delta H^{0,f}_{CaCO_3}$? Explain.

(a) $CaO(s) + CO_2(g) \rightarrow CaCO_3(s)$
(b) $Ca(s) + C(s) + 3/2O_2(g) \rightarrow CaCO_3(s)$

2.9 EXERCISE PROBLEMS

[2.1] Calculate the transformation enthalpy for the isothermal transformation of Ti(β) to Ti(a) if the transformation occurs 200°C below the equilibrium transformation temperature during cooling.

$$Ans:\ \Delta H^{Tr}_{955} = -4227\ \text{J/mol.}$$

[2.2] Derive an expression for the heat of fusion of Si as a function of temperature at 1 atm pressure.

$$Ans:\ \Delta H^f_{T^f}\ \text{(J/mol)} = 51,157 + 1.51T^f$$

$$-0.00117T^{f^2} + 4.56 \times 10^5 / T^f.$$

[2.3] Modified austempering is an isothermal transformation heat treatment used to produce bainite and pearlite (a mixture of ferrite (α) and cementite Fe_3C). The process involves austenitizing the steel to 100% austenite (γ) and then quenching to a temperature slightly below the nose of the "I-T" curve and holding at that temperature until the transformation is complete. Estimate the transformation enthalpy for the transition of a 1050 steel at the austempering temperature. Refer to the I-T diagram for 1050 steel, Appendix E, Figure E.1, to select an austempering temperature. Assume carbon solubility has a negligible effect on the result in either phase. Do not include Fe_3C in the calculations.

$$Ans:\ \Delta H^{Tr}_{773} = -6855\ \text{J/mol.}$$

[2.4] The results of constant volume molar heat capacity calculations for several solids are illustrated in Figure 2.6. For each substance, C_v approaches zero as the absolute temperature approaches zero. In addition, C_v asymptotically approaches a constant value of 6 cal/(mol·°C) as temperature increases. This constant, known as the *Dulong and Petit value*, is best approximated for each solid in a different temperature range. For lead, this temperature range begins at about 180 K and for Cu, at about 500 K.

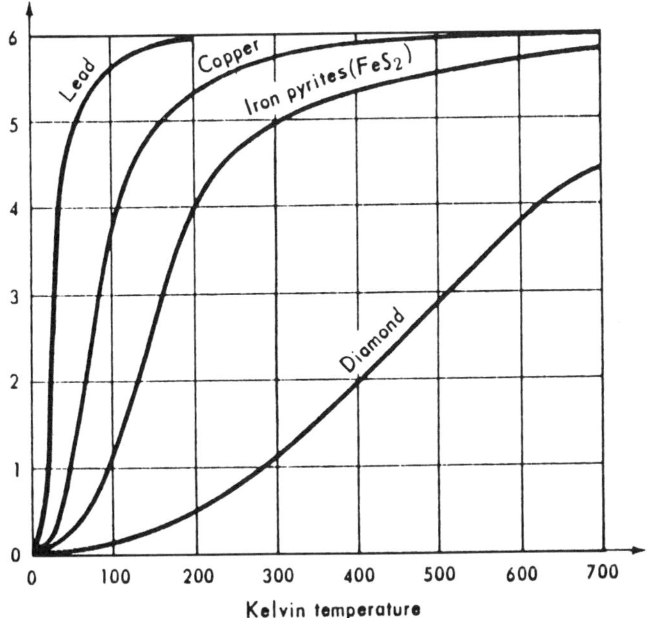

Figure 2.6 Variation of C_v of solids with Kelvin temperature. (From M.W. Zemansky and H.C. Van Ness, 1966, Basic Engineering Thermodynamics, Fig. 11.16. Reprinted by permission of McGraw Hill, Inc., New York.)

Calculate the molar heat transferred to or from the surroundings and the molar internal energy change occurring at constant volume when:
(a) Cu undergoes a temperature increase from 600 to 700 K.

Ans: $Q_v = \Delta U_v = +600$ cal/mol Cu, from the surroundings.
(b) Pb undergoes a temperature decrease from 210 to 180 K.

Ans: $Q_v = \Delta U_v = -180$ cal/mol Pb, to the surroundings.

[2.5] Calculate the enthalpy change at 968 K for the reaction
$$2SiO_2(s) + 2Al_2O_3(s) \rightarrow 2Al_2SiO_5(s).$$
Use the following information (all phases are solids):

(a) $2SiO_2 + 3Al_2O_3 \rightarrow Al_6Si_2O_{13}$ $\qquad \Delta H^0_{968} = -22.6$ kJ/mol

(b) $2Al_2SiO_5 + Al_2O_3 \rightarrow Al_6Si_2O_{13}$ $\qquad \Delta H^0_{968} = -39.4$ kJ/mol

Ans: $\Delta H^0_{968} = 8.4$ kJ/mol SiO_2.

[2.6] Calculate the standard enthalpy of formation of $CaCO_3$ at 298 K in kcal/mol and kJ/mol from the following data:

(a) $CaO(s) + CO_2(g) \rightarrow CaCO_3(s)$ $\qquad \Delta H^0_{298} = -42.85$ kcal/mol

(b) $Ca(s) + 1/2O_2(g) \rightarrow CaO(s)$ $\qquad \Delta H^{0,f}_{298} = -151.50$ kcal/mol

(c) $C(s) + O_2(g) \rightarrow CO_2(g)$ $\qquad$ $\Delta H_{298}^{0,f} = -94.05$ kcal/mol

$\qquad$ *Ans:* $\Delta H_{298}^{0,f} = -288.4$ kcal/mol $= -1207$ kJ/mol.

[2.7] Given $\Delta H_T^{0,f} = -1133$ kJ/mol O_2 for the reaction

$\qquad$ (4/3)Al(s) + O_2(g) = (2/3)Al_2O_3(s),

(a) Show $\Delta H_{298}^{0,f} = -1133$ kJ/mol if $\Delta C_p \approx 0$.

(b) Select the correct alternatives below for $\Delta H_T^{0,f}$.

$\quad$ (1) 4 Al(s) + 3O_2(g) = 2Al_2O_3(s) $\qquad$ $\Delta H_T^{0,f} = -1133$ kJ/mol O_2

$\quad$ (2) 2Al(s) + (3/2)O_2(g) = Al_2O_3(s) $\qquad$ $\Delta H_T^{0,f} = -3399$ kJ/mol Al

$\quad$ (3) 4Al(s) + 3O_2(g) = 2Al_2O_3(s) $\qquad$ $\Delta H_T^{0,f} = -1700$ kJ/mol O_2

$\quad$ (4) 2Al(s) + (3/2)O_2(g) = Al_2O_3(s) $\qquad$ $\Delta H_T^{0,f} = -850$ kJ/mol Al

$\quad$ (5) 2Al(s) + (3/2)O_2(g) = Al_2O_3(s) $\qquad$ $\Delta H_T^{0,f} = -1700$ kJ/mol Al

$\qquad$ *Ans:* (1) and (4).

[2.8] Fifty grams of the mineral ilmenite, $FeTiO_3$, are heated in an insulated furnace and then quenched in a 500 gm copper vessel containing 200 gm of water. The temperature of the water and vessel both rise from 30°C to 65°C. To what temperature was the ilmenite heated prior to quenching?

$\qquad$ *Ans:* $T = 924$°C.

[2.9] Calculate the final temperature and enthalpy change of the mineral galena, PbS, when 253.62 gm of the mineral at 80°C are placed into 58.73 gm of adiabatically contained water initially at a temperature of 27°C. Express the enthalpy change in cal/mol PbS.

$\qquad$ *Ans:* $T = 36$°C, $\Delta H_{PbS} = -499$ cal/mol.

[2.10] Calculate the heat absorbed and standard enthalpy change at (a) 373 K and (b) 373 K < T < 800 K resulting from the dehydration of the mineral serpentine, $Mg_3Si_2O_5(OH)_4$, in the presence of quartz to form talc, $Mg_3Si_4O_{10}(OH)_2$. Use the heat capacity for $H_2O(l)$ in Appendix A, Table A.3A and the heat capacity for $H_2O(g)$ in Appendix A, Table A.3B. The reaction is:

$Mg_3Si_2O_5(OH)_4(s) + 2SiO_2(s) \rightarrow Mg_3Si_4O_{10}(OH)_2(s) + H_2O(g)$.

$\quad$ Serpentine $\qquad$ α-Quartz $\qquad$ Talc

$\qquad$ *Ans:* (a) $Q = \Delta H_{373}^0 = 1787$ J/mol; (b) ΔH_T^0 (J/mol) $= -69.42 \times 10^{-8}T^3 - 2.59T^2 + 4672.2T - 3.70 \times 10^7 T^{-1} - 131,264T^{0.5} - 82,315$.

[2.11] A high pressure mineral reaction occurring in some rocks buried to depths of 30 km or more within the earth involves the decomposition of the mineral albite (Ab) to form the minerals jadeite (Jd) and quartz

(Qz). Calculate the standard enthalpy change at 298 K for the albite decomposition reaction:

$$NaAlSi_3O_8(s) \rightarrow NaAlSi_2O_6(s) + SiO_2(s).$$
$$\text{Ab} \qquad\qquad \text{Jd} \qquad\qquad \alpha\text{–Qz}$$

Ans: $\Delta H^0_{298} = -5$ kJ/mol.

[2.12] Using the standard enthalpy change computed for the reaction in Exercise Problem [2.11], calculate the standard enthalpy change of this reaction at T where 298 K $< T <$ 844 K. No phase changes occur.

Ans: ΔH^0_T (J/mol) = $53.21 \times 10^3 - 238.21T + 7.04 \times 10^{-2}T^2 - 7.57 \times 10^{-6}T^3 + 4.92 \times 10^6 T^{-1} + 8.74 \times 10^3 T^{0.5}$.

[2.13] At 1356 K and 1 atm pressure, the following exothermic reaction occurs:

$$2Cu(l) + (1/2)S_2(g) \rightarrow \gamma\text{-}Cu_2S(s).$$

Calculate the heat evolved and enthalpy of formation of γ-$Cu_2S(s)$ at 1356 K and 1 atm pressure. Use the heat capacity for $S_2(g)$ in Appendix A, Table A.3B. Hint: begin with the reaction

$$2Cu(s) + (1/2)S_2(g) \rightarrow Cu_2S(\alpha).$$

Ans: $Q_{1\text{ atm}} = \Delta H^{0,f}_{1356} = -108$ kJ/mol.

[2.14] Calculate the standard enthalpy of formation of methane, $CH_4(g)$, at 298 K in kJ/mol and kcal/mol from the following information:

(a) $CH_4(g) + 2O_2(g) = CO_2(g) + 2H_2O(g)$ $\Delta H^0_{298} = -890.36$ kJ/mol

(b) $C(s) + O_2(g) = CO_2(g)$ $\qquad\qquad\qquad \Delta H^{0,f}_{298} = -393.51$ kJ/mol

(c) $H_2(g) + (1/2)O_2(g) = H_2O(g)$ $\qquad\quad \Delta H^{0,f}_{298} = -241.814$ kJ/mol

Ans: $\Delta H^{0,f}_{298} = 13.22$ kJ/mol = 3.12 kcal/mol.

[2.15] The enthalpy increment for element A at two temperatures is given as follows:

T (K)	$H^0_T - H^0_{500}$ (J/mol)
1000	1750
700	640

Determine the heat capacity of A which most accurately reflects the data given.

Ans: C^A_p (J/mol) = $2.0 + 2 \times 10^{-3}T$.

[2.16] Derive the formula

$$\Delta H_{T_2} = \Delta H_{T_1} = \int_{T_1}^{T_2} \Delta C_p dT$$

where

(a) ΔH_{T_1} and ΔH_{T_2} are the constant pressure molar enthalpy changes of a reaction in states 1 and 2 respectively.

(b) No phase changes occur between states 1 and 2.

(c) All reaction components exist as one mole in the balanced reaction.

[2.17] Confirm the internal energy change for the transformation of potassium from liquid to vapor phase at 1000 K and 0.753 bar.

$\qquad$ *Ans:* $\Delta u = 1753$ kJ/kg.

[2.18] A 20 kg mass falls 10 meters at constant velocity, thereby causing a paddle wheel to stir 500 g of adiabatically contained water initially at 25°C. Assume for the water a negligible change in volume and that $C_v \approx C_p$.

(a) Find the internal energy change and temperature of the water after the weight has fallen 2,4,6,8, and 10 meters.

(b) Plot both internal energy change and gravitational potential energy versus temperature. What does the plot suggest?

Ans: (a):

Distance Fallen (m)	E_p (J)	$\Delta U' = -W'_{Ad.}$ (J)	$T(°C)$
0.0	1962.0	0.0	25.0
2.0	1570.0	392.4	25.2
4.0	1177.0	784.8	25.4
6.0	784.8	1177.0	25.6
8.0	392.4	1570.0	25.8
10.0	0.0	1962.0	25.9

(b) Linear plot is consistent with $C_v =$ constant.

THE SECOND AND THIRD LAWS OF THERMODYNAMICS

3.1 THERMODYNAMIC EFFICIENCY, THE SECOND LAW AND ENTROPY

Starting with the definition of *thermodynamic efficiency* of a system for the conversion of thermal to mechanical energy

$$\eta = \frac{Q_H - Q_L}{Q_H} \qquad [3\text{-}1]$$

it can easily be shown for a Carnot cycle as illustrated in Example Problem 3-1, that

$$\eta = \frac{T_H - T_L}{T_H} \qquad [3\text{-}2]$$

All steps are reversible, the working substance is an ideal gas, and Q_H and T_H are the heat output from and absolute temperature of a high temperature *heat source* respectively. Q_L and T_L are the heat input to and absolute temperature of a low temperature *heat sink* respectively. The heat source and heat sink are *reservoirs* or masses so large that they can transmit or absorb, respectively, an unlimited amount of heat without experiencing a change in temperature.

It should be noted that the Carnot cycle represents maximum theoretical thermodynamic efficiency. A more realistic expression for Carnot efficiency, which takes into account the ΔT required for heat transfer, is discussed in Exercise Problem [3.17].

Rearranging [3-1] and [3-2] and eliminating η,

$$\frac{Q_H}{T_H} = \frac{Q_L}{T_L} \quad \text{or} \quad \frac{Q_H}{T_H} - \frac{Q_L}{T_L} = 0 \qquad [3\text{-}3]$$

Generalizing [3-3] for smaller and smaller Carnot cycles,

$$\oint \frac{\delta Q^{\text{Rev}}}{T} = 0 \qquad [3\text{-}4]$$

The ratio $\delta Q^{\text{Rev}}/T$ defines the thermodynamic function *entropy, S*, where

$$dS = \frac{\delta Q^{\text{Rev}}}{T} \qquad [3\text{-}5]$$

Mathematically, T is an integrating factor since the inexact differential δQ^{Rev} is made exact by dividing by the absolute temperature. Based on the previous discussion, the *Second Law of Thermodynamics* states that entropy, defined by equation [3-5], is a thermodynamic property.

Example Problem 3-1

Plot the Carnot cycle on a temperature-entropy diagram. Using the definition of entropy [3-5], prove [3-2]. The processes in the Carnot cycle involve: (1) *isentropic* (constant entropy) expansion, (2) isothermal transfer of heat from a high temperature (T_H) reservoir, (3) isentropic compression, and (4) isothermal transfer of heat to a low temperature (T_L) reservoir.

Solution

The Carnot cycle processes are plotted on the *T-S* diagram below and [3-5] is applied.

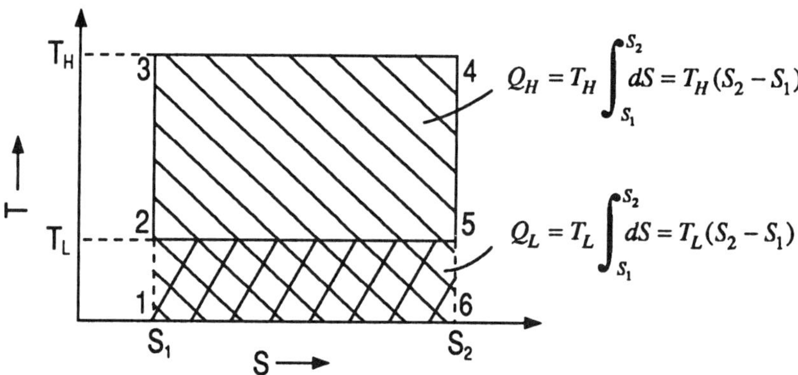

Substituting for Q_H and Q_L in [3-1],

$$\eta = \frac{T_H(S_2 - S_1) - T_L(S_2 - S_1)}{T_H(S_2 - S_1)} = \frac{(T_H - T_L)/T_H}{}.$$

An alternate solution follows from [3-1]:

$$\eta = \frac{Q_H - Q_L}{Q_H} = \frac{\text{Area 2345}}{\text{Area 123456}} = \frac{(T_H - T_L)(S_2 - S_1)}{T_H(S_2 - S_1)} = \frac{(T_H - T_L)/T_H}{}.$$

Now, substituting [1-10] and [3-5] into [2-2],

$$dU = TdS - PdV \qquad\qquad\qquad [3\text{-}6]$$

Writing [2-4] in differential form,

$$dH = dU + PdV + VdP,$$

and substituting [3-6] into this expression,

$$dH = TdS + VdP \qquad\qquad\qquad [3\text{-}7]$$

At constant pressure, [3-7] reduces to

$$dS = \frac{dH_p}{T} \qquad\qquad\qquad [3\text{-}8]$$

Integrating [3-8] at constant temperature,

$$\int_{S_1}^{S_2} dS = (1/T) \int_{H_1}^{H_2} dH_p$$

or

$$\Delta S = \frac{\Delta H_P}{T} \qquad [3\text{-}9]$$

Substituting [2-12] into [3-8] and integrating with respect to temperature,

$$\Delta S = \int_{T_1}^{T_2} \frac{C_p \, dT}{T} \qquad [3\text{-}10]$$

At constant volume, [3-6] reduces to

$$dS = \frac{dU_v}{T} \qquad [3\text{-}11]$$

Integrating [3-11] at constant temperature,

$$\int_{S_1}^{S_2} dS = (1/T) \int_{U_1}^{U_2} dU_v$$

or

$$\Delta S = \frac{\Delta U_v}{T} \qquad [3\text{-}12]$$

Substituting [2-9] into [3-11] and integrating with respect to temperature,

$$\Delta S = \int_{T_1}^{T_2} \frac{C_v \, dT}{T} \qquad [3\text{-}13]$$

3.2 ENTROPY OF REACTION AND THE THIRD LAW

Consider the *allotropic** transformation of tin (Darken and Gurry, 1953, p. 197) at 1 atm pressure. Using TL analysis,

(1) Set Up. A thermodynamic loop is set up as follows:

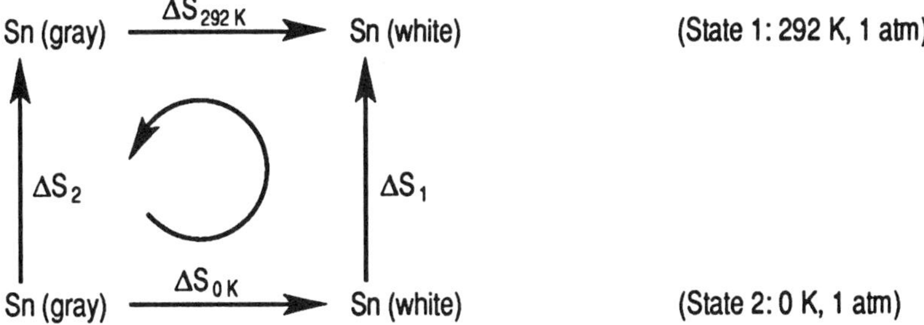

* Allotropes are polymorphic *elements.*

(2) Sum. Summing in the counterclockwise direction,

$$\Sigma\Delta S_{TL} = 0 = \Delta S_{0\,K} + \Delta S_1 - \Delta S_{292\,K} - \Delta S_2.$$

(3) Substitute.[†]

$$\Delta S_1 = \int_0^{292} C_p(\text{white Sn})dT\,/\,T = -38.12\ \text{J/(mol}\cdot\text{K)}$$

$$\Delta S_2 = \int_0^{292} C_p(\text{gray Sn})dT/T = -46.19\ \text{J/(mol}\cdot\text{K)}$$

$$\Delta S_{292} = \Delta H_{292}/T^{Tr} = 2263.5/292 = 7.75\ \text{J/(mol}\cdot\text{K)}.$$

Substituting the above data into $\Sigma\Delta S_{TL} = 0$,

$$0 = \Delta S_{0\,K} - 38.12 - 7.75 + 46.19.$$

(4) Solve.

$$\Delta S_{0\,K} = \underline{-0.32\ \text{J/(mol}\cdot\text{K)}}.$$

Assuming that the calculated value of $\Delta S_{0\,K}$ is within experimental error, it has been shown that $\Delta S_{0\,K}$ is zero. Consequently, it is concluded that the entropies of gray and white tin are the same at 0 K. It follows from the analysis above that the entropies of all substances are the same at 0 K and therefore assumed to be zero at 0 K. This conclusion is stated as the *Third Law of Thermodynamics*: "the entropy of any homogeneous substance which is in complete internal equilibrium may be taken as zero at 0 K" (Darken and Gurry, 1953, p. 193). As will be discussed in Section 3.5, complete order implies the same conclusion. For additional reading concerning the third law, refer to Swalin (1964, p. 51–58), Gaskell (1981, p. 134–145), and DeHoff (1993, p. 32–34). In the context of the previous discussion, standard state entropy values are absolute within the limits of experimental error and are denoted by a superscript 0 in Table 3.1. Since S_{0K}^0 of any homogeneous substance in complete internal equilibrium is zero, the *absolute standard entropy* of any such substance at temperature T can be calculated from [3-10]:

$$\Delta S_T^0 = S_T^0 = \int_0^T \frac{C_p dT}{T}.$$

[†] Below 80 K, C_p was obtained from a combination of Debye's formula and an expression for $C_p - C_v$ given in Exercise Problem [4.12]. Above 80 K, C_p was obtained experimentally.

Table 3.1: Standard State Entropy

State of Aggregation	Pure Elements $P^0 = 1$ atm, T (K)	Pure Compounds $P^0 = 1$ atm, T (K)
Solid (internal equilibrium)	$S^0_{0K} = 0$	$S^0_{0K} = 0$

Values of S^0_{298K} for selected pure solids are found in Table A.1.

The *standard entropy of reaction* is

$$\int_{Reactants}^{Products} dS = \Delta S^0_T$$

or

$$\Delta S^0_T = \sum n S^0_T \text{ (Products)} - \sum n S^0_T \text{ (Reactants)} \qquad [3\text{-}14]$$

Consider the standard state oxidation of element A at a temperature T = 298 K as given by

$$3A + O_2 \rightarrow A_3O_2.$$

The entropy of the reaction is

$$\Delta S^0_{298} = (1)\, S^0_{298,A_3O_2} - (3)\, S^0_{298,A} - (1)\, S^0_{298,O_2}$$

where $S^0_{298,A} \neq 0$, $S^0_{298,O_2} \neq 0$.

Example Problem 3-2

Calculate ΔS^0 at 298 K and 1 atm for the reaction

$$W(s) + O_2(g) = WO_2(s).$$

Solution

Applying [3-14],

$$\Delta S^0_{298} = (1)\, S^0_{298,WO_2} - (1)\, S^0_{298,W} - (1)\, S^0_{298,O_2}$$

$$= 66.95 - 33.48 - 205.10$$

$$= \underline{-171.63 \text{ J/(mol·K)}}.$$

In this example WO_2 is formed from its component elements, all in the most stable form at 298 K and 1 atm. Therefore, $\Delta S^0_{298} = \Delta S^{0,f}_{298}$ where $\Delta S^{0,f}_{298}$ is the *standard entropy of formation* of WO_2.

3.3 TL ANALYSIS: TEMPERATURE DEPENDENCE

To determine an expression for the entropy change of a reaction as a function of temperature, the procedure described in Example Problem 2-4, Section 2.7 is followed.

Example Problem 3-3

Derive an expression for the standard entropy change of the general reaction $n_A A + n_B B \rightarrow n_C C + n_D D$ as a function of temperature at $P^0 = 1$ atm.

Solution

(1) Set Up.

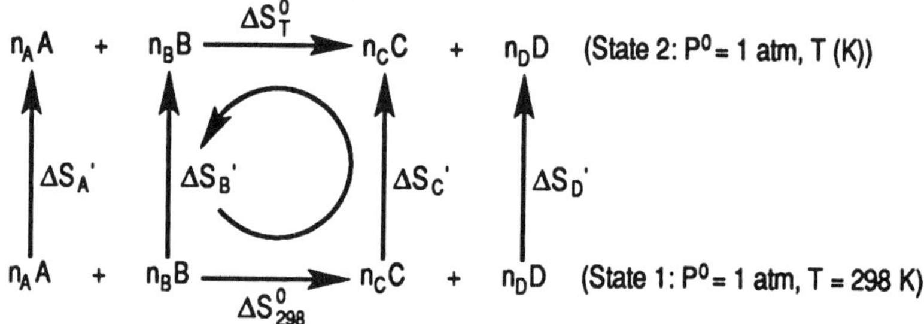

(2) Sum. Summing in the counterclockwise direction,

$$\Sigma \Delta S_{TL} = 0 = \Delta S_{298}^0 + \Delta S_C' + \Delta S_D' - \Delta S_T^0 - \Delta S_A' - \Delta S_B'.$$

(3) Substitute. Applying [3-10] to each reaction component and rearranging,

$$\Delta S_T^0 = \Delta S_{298}^0 + n_C \int_{298}^T \frac{C_p^C dT}{T} + n_D \int_{298}^T \frac{C_p^D dT}{T} - n_A \int_{298}^T \frac{C_p^A dT}{T} - n_B \int_{298}^T \frac{C_p^B dT}{T}$$

$$= \Delta S_{298}^0 + \int_{298}^T \frac{\Delta C_p dT}{T}$$

where ΔC_p is defined by [2-17].

(4) Solve.

$$\Delta S_T^0 = \Delta S_{298}^0 + \int_{298}^T \frac{\Delta C_p dT}{T} \qquad [3\text{-}15]$$

or in general,

$$\Delta S_T^0 = \Delta S_{T_1}^0 + \int_{T_1}^T \frac{\Delta C_p dT}{T} \qquad [3\text{-}16]$$

Approximations similar to [2-19] and [2-20] are

Case I: $\Delta C_p = 0$, hence

$$\Delta S_T^0 = \Delta S_{T_1}^0 \qquad\qquad [3\text{-}17]$$

Case II: $\Delta C_p = a$ (constant), hence

$$\Delta S_T^0 = \Delta S_{T_1}^0 + a\ln(T/T_1) \qquad\qquad [3\text{-}18]$$

Example Problem 3-4

Calculate the entropy of formation of TiC(s), $\Delta S_{\text{TiC}}^{0,f}$, at 1200 K.

Solution

TL analysis is applied in a manner analogous to that of Example Problem 2-5.

(1) **Set Up.** A TL is set up to include the Ti(α) $\rightarrow$Ti(β) phase transformation at 1155 K.

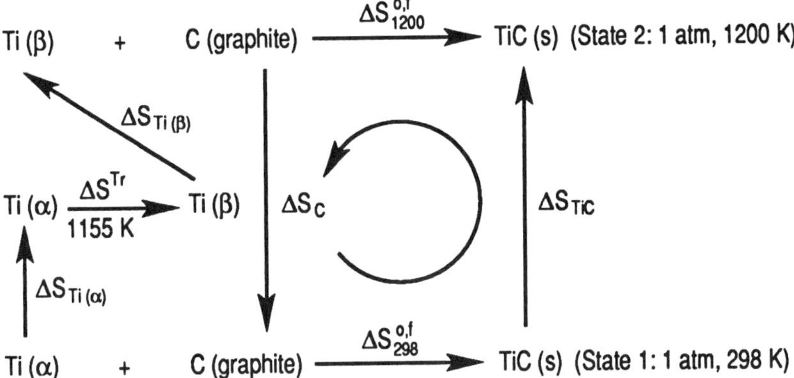

(2) **Sum.** Summing in the counterclockwise direction,

$$\Sigma\Delta S_{\text{TL}} = 0 = \Delta S_{298}^{0,f} + \Delta S_{\text{TiC}} - \Delta S_{1200}^{0,f} + \Delta S_{\text{C}} - \Delta S_{\text{Ti}(\beta)} - \Delta S^{\text{Tr}} - \Delta S_{\text{Ti}(\alpha)}.$$

Rearranging and solving for $\Delta S_{1200}^{0,f}$,

$$\Delta S_{1200}^{0,f} = \Delta S_{298}^{0,f} + \Delta S_{\text{TiC}} + \Delta S_{\text{C}} - \Delta S_{\text{Ti}(\beta)} - \Delta S^{\text{Tr}} - \Delta S_{\text{Ti}(\alpha)}.$$

(3) **Substitute.** $\Delta S_{298}^{0,f}$ is calculated using [3-14] and substituted along with the heat capacity and transformation enthalpy data given below into the expression for $\Delta S_{1200}^{0,f}$.

$$\Delta S_{298}^{0,f} = S_{298,\text{TiC}}^0 - S_{298,\text{Ti}(\alpha)} - S_{298,\text{C}}^0 = -11.96 \text{ J/(mol}\cdot\text{K)}.$$

From heat capacity data,

$$\Delta S_{\text{TiC}} = \int_{298}^{1200} \left(\frac{49.50 + 3.35\times10^{-3}T - 14.98\times10^5 T^{-2}}{T} \right) dT$$

$$\Delta S_C = \int_{1200}^{298} \left(\frac{17.15 + 4.27 \times 10^{-3}T - 8.79 \times 10^5 T^{-2}}{T} \right) dT$$

$$\Delta S_{Ti(\beta)} = \int_{1155}^{1200} \left(\frac{28.91}{T} \right) dT$$

$$\Delta S_{Ti(\alpha)} = \int_{298}^{1155} \left(\frac{22.09 + 10.04 \times 10^{-3}T}{T} \right) dT \,.$$

Substituting ΔH_{Ti}^{Tr} from Example Problem 2-5 into [3-9],
$$\Delta S^{Tr} = \Delta H^{Tr}/T$$
$$= 3473/1155 = 3.01 \; J/(mol \cdot K).$$

(4) Solve. The desired unknown is obtained by integrating and completing calculations. Integration, left to the reader, gives

$$\Delta S_{1200}^{0,f} = \underline{-13.64 \; J/(mol \cdot K)}$$

Note that substitution of $\Delta H_{1200}^{0,f} = -186.7$ kJ/mol (Example Problem 2-5)

and T = 1200 K into [3-9] *cannot* be used to calculate $\Delta S_{1200}^{0,f}$ since the carbide forming reaction is *irreversible*.

3.4 STABILITY CRITERIA: ENTROPY

The *net entropy* change associated with a reaction or process, ΔS_{Net},* is defined by

$$\Delta S_{Net} = \Delta S_{Sys} + \Delta S_{Surr}^{Rev} \tag{3-19}$$

where ΔS_{Sys} denotes the entropy change of the system and ΔS_{Surr}^{Rev} denotes the *reversible* entropy change of the surroundings (see Swalin, 1964, p. 17-20 for additional information regarding ΔS_{Net}). Equation [3-19] can be used to assess whether or not a reaction or process will proceed. Consequently, it assesses the stability of the products under the conditions for which ΔS_{Net} is calculated.

Three *stability criteria* are defined:
(1) If $\Delta S_{Net} = 0$, the reaction or process is at equilibrium;
(2) If $\Delta S_{Net} > 0$, the reaction or process as described is spontaneous;
(3) If $\Delta S_{Net} < 0$, the reaction or process as described is nonspontaneous.

---

* ΔS_{Net} is also referred to as ΔS_{Total} or as $\Delta S_{Universe}$.

The above criteria are applicable under the following conditions:

(a) The system and surroundings are at constant pressure, hence from [2-6], $dH = \delta Q_p$.

(b) The *surroundings* behave reversibly and isothermally even though the reaction is proceeding irreversibly. Thus $\delta Q_{Surr}^{Rev} = -\delta Q_{Sys}$ and from [2-6], $dH_{Surr}^{Rev} = -dH_{Sys}$.

Combining [2-6] and [3-5] with $T = T_{Surr}$,

$$dS_{Surr}^{Rev} = \frac{\delta Q_{Surr}^{Rev}}{T_{Surr}} = \frac{-\delta Q_{Sys}}{T_{Surr}} = \frac{-dH_{Sys}}{T_{Surr}} \qquad [3\text{-}20]$$

and upon integration as in [3-9],

$$\Delta S_{Surr}^{Rev} = \frac{\Delta H_{Surr}^{Rev}}{T_{Surr}} = \frac{-\Delta H_{Sys}}{T_{Surr}} \qquad [3\text{-}21]$$

(c) The temperature of the system and surroundings are not necessarily the same.

Example Problem 3-5

Referring to Example Problem 2-6, calculate the entropy change for the isothermal fusion of 1 gram atom of gold supercooled 230°C below the freezing point at atmospheric pressure (Upadhyaya and Dube, 1977, p. 39). What is the entropy change of the surroundings if the surroundings are at the supercooled temperature? Is fusion spontaneous?

Solution

Since the problem statement involves fusion, not solidification as in Example Problem 2-6, the direction of the reactions in the TL loop below are reversed so that ΔS_{Au}^f can be determined directly.

(1) Set Up.

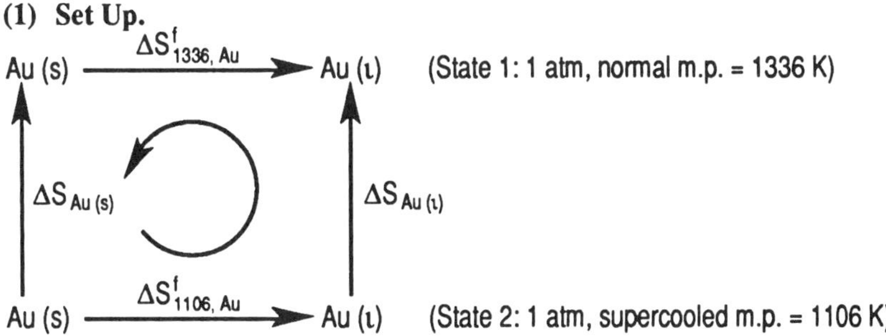

(2) Sum. Summing in the counterclockwise direction,

$$\Sigma \Delta S_{TL} = 0 = \Delta S^f_{1106, Au} + \Delta S_{Au(l)} - \Delta S^f_{1336, Au} - \Delta S_{Au(s)}.$$

Rearranging,

$$\Delta S^f_{1106,\text{Au}} = -\Delta S_{\text{Au}(l)} + \Delta S^f_{1336,\text{Au}} + \Delta S_{\text{Au}(s)}.$$

(3) Substitute. Inserting heat capacity and transformation enthalpy data,

$$\Delta S^f_{1106,\text{Au}} = -\int_{1106}^{1336} \frac{29.29\,dT}{T} + \frac{\Delta H^f_{1336,\text{Au}}}{T^f} + \int_{1106}^{1336}\left(\frac{23.68 + 5.19\times10^{-3}T}{T}\right)dT.$$

(4) Solve. Integrating,

$$\Delta S^f_{1106,\text{Au}}\ [\text{J/(mol}\cdot\text{K)}] = -29.29\ln(1336/1106) + 9.55 + 23.68\times\ln(1336/1106)$$

$$+ 5.19\times10^{-3}\times(1336-1106)$$

$$= -5.53 + 9.55 + 5.67$$

which reduces to

$$\Delta S^f_{1106,\text{Au}} = \underline{9.69\ \text{J/(mol}\cdot\text{K)}}.$$

The TL analysis above refers to the system. From [3-21] and Example Problem 2-6, the entropy of the surroundings is

$$\Delta S^{\text{Rev}}_{\text{Surr}} = \frac{-\Delta H^f_{1106,\text{Au Sys}}}{T_{\text{Surr}}} = \frac{-12{,}930}{1106}\ \text{J/(mol}\cdot\text{K)}$$

$$= \underline{-11.69\ \text{J/(mol}\cdot\text{K)}}.$$

From [3-19], $\Delta S_{\text{Net}} = (9.69 - 11.69) = \underline{-2.00\ \text{J/(mol}\cdot\text{K)}}$.

Since $\Delta S_{\text{Net}} < O$, fusion is *not* spontaneous at 1106 K.

3.5 ALTERNATE DEFINITION OF ENTROPY

An alternate equation defines entropy according to

$$S' = \text{k}\ln(W) \qquad\qquad [3\text{-}22]$$

where W = number of microstates/macrostate and k is *Boltzmann's constant* (R/N_{Av}). S' reflects the degree of *randomness* or *disorder* of a system.

In the context of crystalline materials at constant temperature, W corresponds to a configurational or structural change as reactants combine to form compounds or solution mixtures. Consider a physical model of a substitutional solid solution made up of atoms of A and B. Suppose atoms of the same kind are indistinguishable from one another but that A atoms are distinguishable from B atoms. It can then be shown that,

$$W = \frac{(N_{\text{A}} + N_{\text{B}})!}{N_{\text{A}}! \times N_{\text{B}}!}$$

where N_{A} and N_{B} are numbers of A atoms and B atoms respectively. Substituting this expression into [3-22] and expanding by Stirling's approximation,

$$\ln(X!) = X\ln(X) - X,$$

$$S^m = -R[X_{\text{A}}\ln(X_{\text{A}}) + X_{\text{B}}\ln(X_{\text{B}})] \qquad\qquad [3\text{-}23]$$

where S^m is the molar entropy of mixing, and X_A and X_B are the mole fractions of atoms A and B respectively (see SYMBOLS FOR MOLAR PROPERTIES: CHARACTERIZED BY VOLUME). For an equimolar solution, $X_A = X_B = 0.5$; $S^m = -8.3144 \times [0.5 \ln(0.5) + 0.5 \ln(0.5)] = 5.76$ J/(mol·K). Equation [3-23] will be derived using the classical approach in Chapter 6, Section 6.4.

If A and B atoms are indistinguishable from one another, it can be shown that

$$W = \frac{(N_A + N_B)!}{N_A! \times N_B!} = 1.0 \text{ and } S^m = 0.$$

In such a case, the mixture of A and B atoms is completely ordered.

In general, entropy increases as the degree of randomness increases. This behavior is noted, for example, in the transition from solid to liquid (melting), liquid to gas (vaporization), and solid to gas (sublimation). At 0 K, complete order implies that entropy of a homogeneous substance is zero.

3.6 DISCUSSION QUESTIONS

(3.1) Entropy is related to the degree of randomness or disorder in a system (Section 3.5). Give three examples that illustrate a change in randomness and state how each change effects entropy.

(3.2) Consider the chemical reaction $n_A A + n_B B \rightarrow A n_A B n_B$. If $\Delta S^0 > 0$ for this reaction, can it be concluded that the reaction is spontaneous?

(3.3) Assuming the reaction in (3.2) is spontaneous, does

$$\Delta S_T^0 = \Delta H_T^0 \text{ (reaction)}/T \text{ (reaction)}?$$

(3.4) Is S_{298}^0 of an internally equilibrated homogeneous element or compound zero?

(3.5) Two phase transformations occur at a temperature T and pressure P. The phases involved are A and A′, and the corresponding entropy changes are as follows:
(1) A → A′ ΔS_1
(2) A → A′ ΔS_2
How is ΔS_1 related to ΔS_2?

(3.6) For which mineral reaction below is $\Delta S_T^0 = \Delta S_{T,\text{Anorthite}}^{0,f}$?
(a) $SiO_2(s) + 2Al_2SiO_5(s) + Ca_3Al_2(SiO_4)_3(s) \rightarrow 3CaAl_2Si_2O_8(s)$
    Quartz   Kyanite   Grossular   Anorthite
(b) $3Ca(s) + 6Al(s) + 6Si(s) + 12O_2(g) \rightarrow 3CaAl_2Si_2O_8(s)$

3.7 EXERCISE PROBLEMS

[3.1] Prove that PV^γ is constant for an ideal gas. The process is adiabatic. Let $\gamma = C_p/C_v$.

[3.2] Compute the standard entropy of formation for the reaction $2H_2(g) + S_2(g) \rightarrow 2H_2S(g)$ from 300 to 800 K in 100 K increments. Plot the entropy as a function of temperature. What is the significance of the plot?

> *Ans:* $\Delta S^{0,f}$ (J/K) = –78.07 at 300 K, –84.42 at 400 K, –89.00 at 500 K, –92.36 at 600 K, –94.83 at 700 K, and –96.62 at 800 K. The plot shows that $\Delta S^{0,f}$ (reaction) decreases as the temperature increases (since $V_{Product} < V_{Reactants}$) and that the volume effect on $\Delta S^{0,f}$ becomes less dominant.

[3.3] Compute the standard molar volume change for the reaction in Exercise Problem [3.2] from 300 K to 800 K in 100 degree increments. Plot the entropy change of the reaction as a function of molar volume change. What is the significance of the plot?

> *Ans:* ΔV^0 (cm³/mol) = –24,617 at 300 K, –32,823 at 400 K, –41,029 at 500 K, –49,234 at 600 K, –57,440 at 700 K, and –65,646 at 800 K. The plot shows that $\Delta S^{0,f}$ decreases as ΔV^0 decreases. This is consistent with the reduction in volume associated with the reaction as it proceeds in the direction indicated.

[3.4] Use the entropy criteria to prove that liquid tantalum will solidify if the surroundings are at 20°C. Assume that the melt does not supercool.

> *Ans:* ΔS_{Net} = 76.36 J/(mol·K) > 0, hence Ta will solidify.

[3.5] One mole of low carbon steel at 900 K is quenched and held in contact with a large reservoir of liquid lead at 700 K.

(a) What is ΔS for the steel, for the reservoir, and for the universe (ΔS_{Net})? Use $C_p^{Fe(\alpha)}$ and assume that the reservoir is sufficiently large to maintain constant temperature.

> *Ans:* ΔS_{Steel} = –9.36 J/(mol·K), ΔS_{Res} = 10.67 J/(mol·K), and ΔS_{Net} = 1.32 J/(mol·K).

(b) If the steel had been quenched to an intermediate temperature of 800 K, then subsequently quenched to 700 K, what would ΔS_{Net} be for such a process? Note that during the quenching process the steel is placed in contact with reservoirs at consecutively lower temperatures.

> *Ans:* ΔS_{Net} = 0.63 J/(mol·K).

(c) Explain how "equilibrium quenching" could be approached.

> *Ans:* ΔS_{Net} = 0.

[3.6] Calculate ΔS_{Sys}, ΔS_{Surr}, and ΔS_{Net} for the isothermal solidification of lithium at 453 K and 1 atm pressure. Assume $T_{Sys} = T_{Surr}$.

> *Ans:* ΔS_{Sys} = –6.47 J/(mol·K), ΔS_{Surr} = 6.47 J/(mol·K), and ΔS_{Net} = 0.

[3.7] Calculate the entropy change of: (a) the system, (b) the surroundings, and (c) the universe (ΔS_{Net}) for the isothermal solidification of 1 gram atom of Al supercooled 130°C below its normal freezing point. Assume $\Delta C_p = 0$ over the supercooled interval. Is the process spontaneous?

Ans: $\Delta S_{Sys} = -11.21$ J/(mol·K), $\Delta S_{Surr} = 13.03$ J/(mol·K), ΔS_{Net}
$= 1.82$ J/(mol·K) $> 0 \Rightarrow$ spontaneous.

[3.8] Repeat Exercise Problem [3.7] using heat capacity data from Appendix A, Table A.3A. What is the percent error in ΔS_{Net} assuming $\Delta C_p = 0$?

Ans: $\Delta S_{Sys} = -11.53$ J/(mol·K), $\Delta S_{Surr} = 15.11$ J/(mol·K), ΔS_{Net}
$= 3.58$ J/(mol·K) $> 0 \Rightarrow$ spontaneous. Error $\approx 100\%$.

[3.9] Calculate the entropy change for the vaporization of potassium at 950 K and 0.447 bar. Illustrate calculations using two methods.

Ans: $\Delta s^{l \rightarrow v} \approx 2.08$ kJ/(kg·K).

[3.10] Use the entropy criteria to confirm that liquid and potassium vapor are in equilibrium at 950 K and 0.447 bar.

Ans: $\Delta S_{Net} \approx 0$.

[3.11] A single crystal of the amphibole tremolite at 298 K undergoes a volume expansion of 105.5×10^{-3} cm³/mol during heating at $P^0 = 1$ atm.
(a)Calculate the temperature shift associated with the volume change. Assume $\alpha V_{1atm, 298 K}$ is constant over the temperature interval.

Ans: $\Delta T = 12.49$ K.

(b)Use the result from (a) to calculate the corresponding change in entropy.

Ans: $\Delta S = 2.73 \times 10^{-2}$ kJ/(mol·K).

[3.12] A quartz sandstone with calcite cement is intruded by molten igneous rock as shown below.

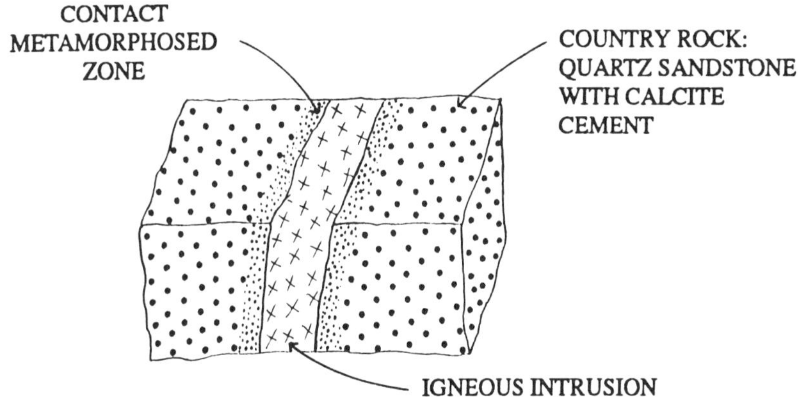

The country rock (sandstone) adjacent to the intrusion undergoes "contact metamorphism" as a consequence of heat transfer from the intrusion. The contact metamorphic reaction is:

$$SiO_2(s) + CaCO_3(s) \rightarrow CaSiO_3(s) + CO_2(g).$$

α-Quartz   Calcite   Wollastonite

Calculate ΔS^0 for this reaction at 565°C.

Ans: $\Delta S^0_{565\,°C} = 154$ J/(mol·K).

[3.13] Calculate the standard entropy of formation for the reaction in Exercise Problem [2.13] at 1356 K.

Ans: $\Delta S^{0,f}_{1356} = 122.2$ J/(mol·K).

[3.14] Calculate the standard entropy change for the reaction in Exercise Problem [2.10] at temperature T where 373 K < T < 800 K. No phase changes occur.

Ans: ΔS^0_T [J/(mol·K)] $= -771.17 + 4672.2 \ln(T) - 5.17T -$
$10.41 \times 10^{-7}T^2 + 131.26 \times 10^3 T^{-0.5} - 389.32 \times 10^3 T^{-2}$.

[3.15] Calculate the standard entropy change for the reaction in Exercise Problem [2.12] at temperature T where 298 K < T < 844 K.

Ans: ΔS^0_T [J/(mol·K)] $= 1762.2 - 238.21\ln(T) + 140.75 \times 10^{-3}T$
$- 113.61 \times 10^{-7}T^2 - 87.38 \times 10^2 T^{-0.5} + 24.60 \times 10^5 T^{-2}$.

[3.16] Consider the hypothetical power cycle represented below.
(a) Calculate actual and Carnot efficiencies for the process if $T_2 = 600$ K and $T_1 = 300$ K.
 Ans: $\eta = 33\%$, $\eta_{Car} = 50\%$.
(b) If a new alloy permits raising the maximum operating temperature to 750 K, calculate the resulting efficiencies.
 Ans: $\eta = 43\%$, $\eta_{Car} = 60\%$.

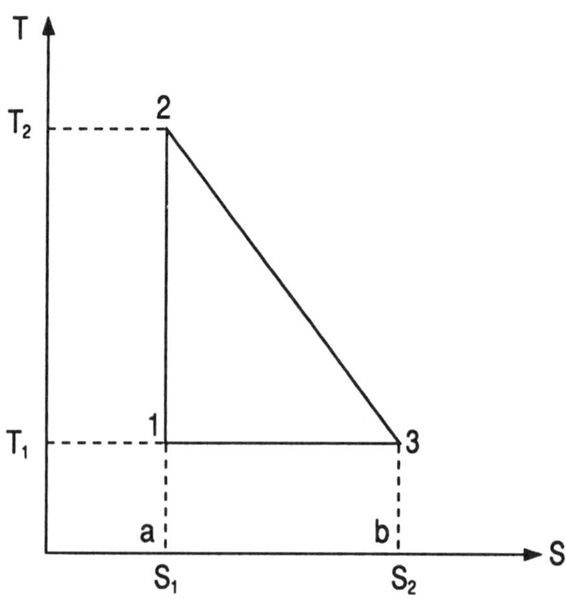

[3.17] The Carnot cycle efficiency is often used as a basis for estimating the effect of increasing operating input temperature and lowering rejection temperature on the efficiency of a power generating process. Curzon and Ahlborn (1975, p. 22-24) have shown that to achieve theoretical Carnot efficiency, heat transfer must take place through an infinitely small temperature differential, which in turn, requires infinite time. They derived the following modification to Carnot efficiency, which takes into account the necessary temperature differential that must exist across the high and low temperature heat transfer boundaries, if the cycle is going to be completed in a reasonable time interval:

$$\eta_{C.A.} = 1 - \sqrt{T_L / T_H} .$$

Using the data given in Exercise Problem [3.16], calculate $\eta_{C.A.}$ and compare with η and η_{Car} obtained in the same problem.

Ans: (a) $\eta_{C.A.} = 29.3\%$, (b) $\eta_{C.A.} = 36.8\%$. The predicted efficiencies are closer to the actual values in [3.16]. On a comparative basis, the $\eta_{C.A.}$ equation provides a stronger case for alloy modification.

[3.18] A small amount of $H_2S(g)$ in volcanic gases promotes intense chemical weathering of rocks in the vicinity of a volcanic vent. Calculate the standard entropy of formation associated with the production of $H_2S(g)$ from its elements at 600 K and 1 atm in kJ/(mol·K) and cal/(mol·K).

(a) $3H_2(g) + SO_2(g) \rightarrow 2H_2O(g) + H_2S(g)$

$\Delta S_{600}^0 = -75.44$ J/(mol·K)

(b) $SO_2(g) \rightarrow (1/2)S_2(g) + O_2(g)$

$\Delta S_{600}^0 = 73.26$ J/(mol·K)

(c) $H_2(g) + (1/2)O_2(g) \rightarrow H_2O(g)$

$\Delta S_{600}^{0,f} = -51.30$ J/(mol·K)

Ans: $\Delta S_{600}^0 = -46.1 \times 10^{-3}$ kJ/(mol·K) $= -11.02$ cal/(mol·K).

[3.19] Show that $H_2S(g)$ spontaneously forms at 600 K. Refer to Exercise Problem [3.18]. Assume the surroundings are at 600 K. Use the heat capacity for $S_2(g)$ in Appendix A, Table A.3B.

Ans: $\Delta S_{Net} = 100.5$ J/(mol·K) $> 0 \Rightarrow$ spontaneous.

GIBBS FREE ENERGY

4.1 STABILITY CRITERIA

For a spontaneous process or reaction it is noted from Section 3.4 that ΔS_{Net} is positive. Expressing [3-19] in differential form for such a process

$$dS_{Net} = dS_{Sys} + dS_{Surr}^{Rev} > 0 \qquad [4\text{-}1]$$

and substituting [3-20] into [4-1], Swalin (1964, p. 25–26) shows that

$$dS_{Net} = dS_{Sys} - dH_{Sys}/T > 0.$$

The designation Sys is dropped since the inequality refers to the *system*, hence $TdS - dH > 0$. Reversing the inequality sign,

$$dH - TdS < 0 \qquad [4\text{-}2]$$

According to [4-2], $dH - TdS$ is less than zero for a *spontaneous* reaction. If the reaction is *nonspontaneous*,

$$dH - TdS > 0 \qquad [4\text{-}3]$$

At equilibrium,

$$dH - TdS = 0 \qquad [4\text{-}4]$$

The expression $dH - TdS$ is related to the *Gibbs free energy* function G as follows: Starting with the definition

$$G = H - TS \qquad [4\text{-}5]$$

and taking its differential,

$$dG = dH - TdS - SdT \qquad [4\text{-}6]$$

At constant temperature, [4-6] reduces to

$$dG = dH - TdS \qquad [4\text{-}7]$$

Substituting [4-7] into [4-2], [4-3], and [4-4] and integrating, stability criteria are defined in terms of Gibbs free energy:

(1) If $\Delta G = 0$, the reaction is at equilibrium;
(2) If $\Delta G < 0$, the reaction is spontaneous in the direction specified;
(3) If $\Delta G > 0$, the reaction is nonspontaneous in the direction specified.

The Gibbs free energy provides a chemical reaction stability criteria that is based only on the properties of the *system* at constant pressure and temperature.

Example Problem 4-1

Referring to the fusion of supercooled gold, Example Problem 3-5, determine whether or not fusion is spontaneous. Use the Gibbs free energy criteria. Is the result in agreement with entropy criteria?

Solution

Integrating [4-7] at constant temperature,

$$\Delta G_T = \Delta H_T - T\Delta S_T \qquad [4\text{-}8]$$

From Example Problems 2-6 and 3-5, $\Delta H_{1106,Au}^{f} = 12{,}930$ J/mol and $\Delta S_{1106,Au}^{f} = 9.69$ J/(mol·K), hence

$$\Delta G^f_{1106,\,Au} = 12,930 - 1106(9.69) = 12,930 - 10,717$$
$$= \underline{2,213 \text{ J/mol.}}$$

Since $\Delta G^f_{1106,\,Au} > 0$, fusion is *not* spontaneous at 1106 K. This agrees with the answer in Example Problem 3-5.

4.2 STANDARD STATES: GIBBS FREE ENERGY OF FORMATION AND REACTION

The standard state Gibbs free energy of compounds and elements, ΔG^0_T, is chosen at a reference pressure of 1 atm (or 1 bar) and the state of aggregation (solid, liquid, or gas) defined at the temperature of interest. For pure compounds $\Delta G^0_T = \Delta G^{0,f}_T$, the *Gibbs free energy of formation* of the compound. $\Delta G^{0,f}_T$ is calculated from the reaction involving formation of the compound form its constituent *elements*. This is consistent with the definition of enthalpy of formation in that Gibbs free energy of formation may be calculated at any temperature at which the compound and its component elements are in the *most stable* configuration. For pure *elements* in the *most stable* configuration, standard state Gibbs free energy is normally assigned the arbitrary value $\Delta G^0_T = 0$. Table 4.1 summarizes the standard free energy states for elements and compounds.

Table 4.1: Standard State Gibbs Free Energy

State of Aggregation	Pure Elements $P^0 = 1$ atm*, Stable at T (K)	Pure Compounds $P^0 = 1$ atm*, Stable at T (K)
Solid	$\Delta G^0_T = 0$	$\Delta G^0_T = \Delta G^{0,f}_T$
Liquid	$\Delta G^0_T = 0$	$\Delta G^0_T = \Delta G^{0,f}_T$
Gas (ideal)	$\Delta G^0_T = 0$	$\Delta G^0_T = \Delta G^{0,f}_T$

* See footnote about standard pressure: Table 2.1.

As in the case of enthalpy, only changes in Gibbs free energy are computed. Hence for a chemical reaction occurring at 1 atm pressure and $T = T$(K),

$$\Delta G^0_T = \sum n\Delta G^{0,f}_T \text{ (Products)} - \sum n\Delta G^{0,f}_T \text{ (Reactants)} \qquad [4\text{-}9]$$

where ΔG^0_T now represents the *standard Gibbs free energy of the reaction* and n is the number of moles of each reactant or product in a *balanced* chemical reaction. Note the analogy between [4-9] and [2-14].

Suppose element A oxidizes at 298 K and 1 atm to form a hypothetical oxide A_2O according to

$$2A(s) + 1/2O_2(g) \rightarrow A_2O(s).$$

From [4-9] and Table 4.1,

$$\Delta G^0_{298} = (1)\Delta G^{0,f}_{298,A_2O(s)} - 2\Delta G^0_{298,A(s)} - (1/2)\Delta G^0_{298,O_2(g)}$$

or

$$\Delta G^0_{298} = \Delta G^{0,f}_{298,A_2O(s)}.$$

Note, as mentioned in Chapter 2 (Section 2.6), that the superscript f applies to *compounds* but not to elements.

4.3 METHODS FOR DETERMINATION OF ΔG^0_T

Tables, Charts, and Computer Programs

The choice of source materials for thermodynamic data (tables, charts, or computer programs) is often determined by the accuracy required. For the purposes of this book, adequate data is given in the tables in Appendix A for solving either Example or Exercise Problems.

Assuming $\Delta C_p \approx 0$ and substituting equations [2-19] and [3-17] into [4-8],

$$\Delta G^0_T = \Delta H^0_{T_1} - T\Delta S^0_{T_1} \qquad [4\text{-}10]$$

For $T_1 = 298$ K,

$$\Delta G^0_T = \Delta H^0_{298} - T\Delta S^0_{298} \qquad [4\text{-}11]$$

Here, only the standard enthalpy and entropy of reaction at 298 K are needed to compute ΔG^0_T, no C_p data is required. Often, the data at 298 K is the only information available, hence [4-11] is the only alternative.

For many materials, C_p is known as a function of temperature. It is common in such cases for $\Delta C_p = a$ (constant), hence substituting [2-20] and [3-18] into [4-8] gives

$$\Delta G^0_T = \Delta H^0_{298} + a(T - 298 \text{ K}) - T\left[\Delta S^0_{298} + a\ln(T/298K)\right] \qquad [4\text{-}12]$$

Collecting constants,

$$\Delta G^0_T = A + BT\log(T) + CT \qquad [4\text{-}13]$$

The constants A, B, and C are listed in Appendix A, Table A.4, for selected reactions. Note that log is $\log_{10}$ in the table. In the remainder of this book, the letter f for formation will be dropped in order to simplify notation and avoid confusion with the letter f for fusion.

Ellingham Diagrams

Ellingham diagrams (Ellingham, 1944) are plots of standard Gibbs free energy, ΔG^0_T, versus temperature T. Each plot is *approximated* by a straight line for which the linear equation is derived by:

(1) Assuming $\Delta C_p \approx 0$ in the temperature interval for which no reaction components change phase;

(2) Application of [4-10] where $-\Delta S^0_{T_1}$ is the slope (also see [4-27]) and $\Delta H^0_{T_1}$ is the ordinate intercept.

Consider, for example, the reaction $2Mg + O_2 = 2MgO$ shown in Figure 4.1 (taken from Figure E.2). Two breaks in slope appear on this diagram, one at 923 K corresponding to the melting point of Mg, and the other at 1378 K corresponding to the boiling point of Mg. Equation [4-11] is applicable to the linear segment between 298 K (arbitrary lower limit) and the melting point of Mg at 923 K. Between the melting and boiling points of Mg, ΔH^f and ΔS^f are incorporated into Hess's law as follows:

(a) $2Mg(s) + O_2(g) = 2MgO(s)$ $\qquad \Delta G_T^0 = \Delta H_{298}^0 - T\Delta S_{298}^0$

(b) $2Mg(s) = 2Mg(l)$ $\qquad 2\Delta G_{923}^f = 2\Delta H_{923}^f - 2T\Delta S_{923}^f$

(c) Reaction (a) – Reaction (b):
$\quad\;\; 2Mg(l) + O_2(g) = 2MgO(s)$

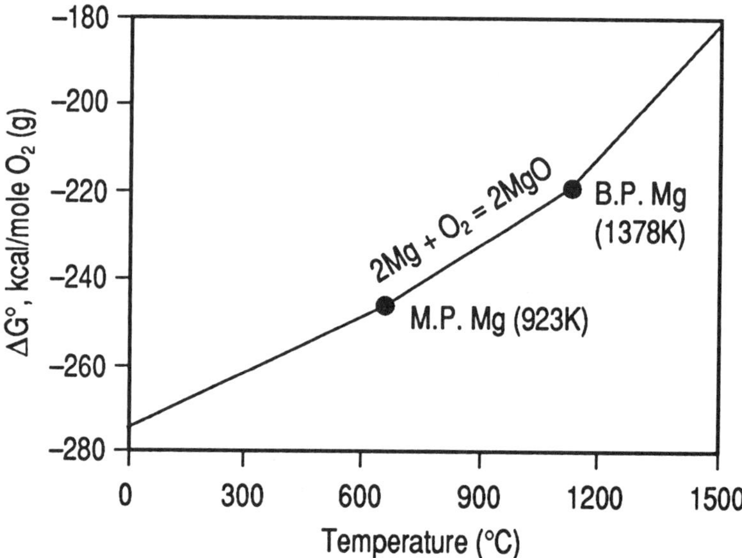

Figure 4.1 Ellingham diagram for the standard state oxidation of magnesium.

ΔG_T^0 for reaction (c) is given by

$$\Delta G_T^0 = \Delta H_{298}^0 - 2\Delta H_{923}^f - T(\Delta S_{298}^0 - 2\Delta S_{923}^f).$$

This expression is applicable to the nearly linear segment between 923 K and 1378 K in Figure 4.1. Since the enthalpy and entropy of fusion are positive, the entropy of reaction (c) is more negative than that of reaction (a). From [4-27], the slope of the reaction line for each segment is

$$\left(\partial \Delta G / \partial T\right)_p = -\Delta S,$$

hence the slope increases at the transformation point when Mg melts. Note that the change in slope is much greater for boiling than for melting because the entropy of vaporization is greater than the entropy of fusion. Ellingham diagrams for metallic oxides and sulfides are shown in Appendix E, Figures

E.2 and E.3 respectively. Particular attention should be given to the notation in the box inserts on each figure. Note that for computation, T (°C) in these figures is converted to T (K).

Example Problem 4-2

Calculate $\Delta G^0_{298 \text{ K}}$ for the reaction

$$Ni + 1/2O_2 \rightarrow NiO.$$

(a) Assume $\Delta C_p = 0$, (b) assume ΔC_p is constant, (c) use complete C_p expressions, and (d) use the Ellingham diagram for oxides.

Solution

(a) **Method 1:** Assuming $\Delta C_p = 0$,* [4-11] is combined with data in Table A.1:

$$\Delta G^0_{298} = \Delta H^0_{298} - (298)\Delta S^0_{298} \Rightarrow$$

$$\Delta G^0_{298} = -240,600 - (298)[38.08 - 29.79 - 0.5(205.10)]$$

$$= \underline{-212,510 \text{ J/mol.}}$$

(b) **Method 2:** Assuming $\Delta C_p = $ constant, use Table A.4.

$$\Delta G^0_T = A + BT\log(T) + CT$$

$$\Delta G^0_{298} = -244,580 + 98.54(298) = \underline{-215,210 \text{ J/mol.}}$$

(c) **Method 3:** Using complete C_p expressions as in [2-18] and substituting [2-15] and [3-15] into [4-8], the result is identical to [4-11] at $T = 298$ K. Hence,

$$\Delta G^0_{298} = \underline{-212,510 \text{ J/mol.}}$$

as in Method 1. However, when $T \neq 298$ K, ΔG^0_T is calculated by using the above substitutions and integrating. For most engineering applications, calculation of ΔG^0_T using the shorter techniques of either Method 1 or Method 2 above is adequate.

(d) **Method 4:** Referring to Figure 4.2 (taken from Figure E.2),

$$\Delta G^0_{298} = -(106,000/2)(4.184) = \underline{-221,750 \text{ J/mol.}}$$

---

* [4-11] is rigorous at T = 298 K, since ΔC_p does not enter into the calculation.

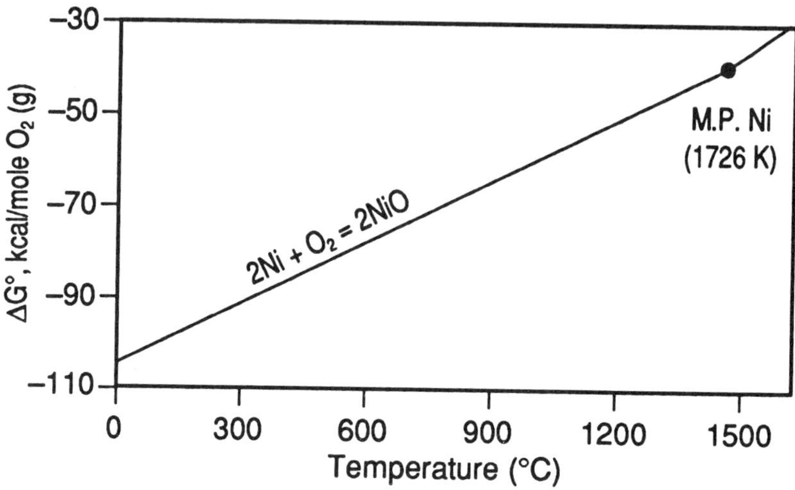

Figure 4.2 Ellingham diagram for the standard state oxidation of nickel.

Referring to Figure E.2, many oxides are more stable than NiO because they appear below NiO and exhibit a more negative ΔG^0. Since all of the reactions on the diagram are based on one mole of O_2, the stabilities of the oxides are directly comparable. For example, carbon is a reducing agent for iron above about 750°C since CO is more stable than FeO. On the other hand, TiO_2 is more stable than CO up to near the melting point of Ti.

Example Problem 4-3

Which metal is more stable in water vapor (superheated steam) at 1000°C, Cr or Ni? Show calculations.

Solution

Referring to Figure 4.3 (taken from Figure E.2), the reaction lines for
$$2Ni(s) + O_2(g) = 2NiO(s)$$
and
$$4/3Cr(s) + O_2(g) = 2/3Cr_2O_3(s)$$
are, respectively, above and below the reaction line for $2H_2(g) + O_2(g) = 2H_2O(g)$. Therefore, $Cr_2O_3(s)$ is more stable than $NiO(s)$ at 1000°C. In terms of the elements, Ni is more stable than Cr in superheated steam at 1000°C.

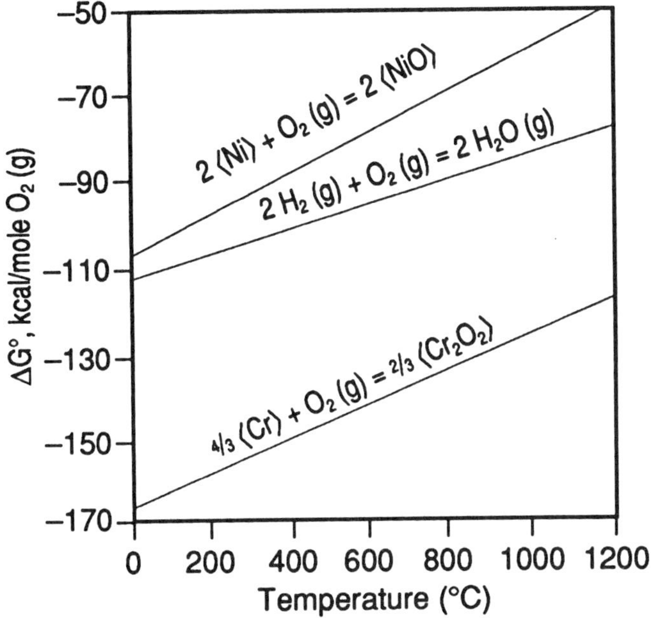

Figure 4.3 Ellingham diagram for the standard state oxidation of nickel, chromium, and hydrogen.

To illustrate numerically, data are read directly from the Ellingham diagram:

(1) $2Ni(s) + O_2(g) = 2NiO(s)$ $\Delta G^0_{1273} = -60$ kcal/mol

(2) $2H_2(g) + O_2(g) = 2H_2O(g)$ $\Delta G^0_{1273} = -86$ kcal/mol

(3) $4/3Cr(s) + O_2(g) = 2/3Cr_2O_3(s)$ $\Delta G^0_{1273} \approx -127$ kcal/mol

(4)   Reaction (1) – Reaction (2):

$2Ni(s) + 2H_2O(g) = 2NiO(s) + 2H_2(g)$ $\Delta G^0_{1273} = \underline{26 \text{ kcal/mol}}$.

Since $\Delta G^0_{1273} > 0$, Ni will tend to deoxidize in steam at 1000°C.

(5)   Reaction (3) – Reaction (2):

$4/3Cr(s) + 2H_2O(g) = 2/3Cr_2O_3(s) + 2H_2(g)$ $\Delta G^0_{1273} = \underline{-41 \text{ kcal/mol}}$.

Since $\Delta G^0_{1273} < 0$, Cr will tend to oxidize in steam at 1000°C. The computations in (4) and (5) confirm the conclusions drawn from the relative positions of the Ni(s) and Cr(s) oxidation lines shown in the diagram.

Example Problem 4-4

Determine the standard enthalpy and entropy change for the reaction

$$Mg + (1/2)O_2(g) \rightarrow MgO(s),$$

using Figure E.2 at temperatures up to the melting point of Mg. Compare the results with the data in Table A.1.

Solution
Arbitrarily selecting two points on the diagram,

(a) $\Delta G^0_{473} = (1/2)(-268 \text{ kcal/mol})(4.184 \text{ J/cal}) = -561 \text{ kJ/mol};$

(b) $\Delta G^0_{923} = (1/2)(-245 \text{ kcal/mol})(4.184 \text{ J/cal}) = -513 \text{ kJ/mol}.$

Substituting into [4-11],

(1) $-561 \text{ kJ/mol} = \Delta H^0_{298} - (473)\,\Delta S^0_{298};$

(2) $-513 \text{ kJ/mol} = \Delta H^0_{298} - (923)\,\Delta S^0_{298}$

Solving (1) and (2) simultaneously,

$$\Delta H^0_{298} = \underline{-612 \text{ kJ/mol}};$$

$$\Delta S^0_{298} = \underline{-0.107 \text{ kJ/(mol} \cdot \text{K)}}.$$

From Table A.1,

$$\Delta H^0_{298} = \underline{-601.3 \text{ kJ/mol}};$$

$$\Delta S^0_{298} = (27.41 - 32.51 - 205.02/2)/1000$$
$$= \underline{-0.108 \text{ kJ/(mol} \cdot \text{K)}}.$$

The two data sets above compare favorably considering the assumptions inherent in [4-11].

Example Problem 4-5

Examine the reaction $2Mg + O_2 = 2MgO$ in Figure 4.1 (or E.2) and explain why the slope increases at 1105°C. Examine the reaction $2Pb + O_2 = 2PbO$ in Figure E.2 and explain why the slope decreases at 1480°C.

Solution
The standard entropy change for the reaction $2Mg(l) + O_2(g) = 2MgO(s)$ is

$$\Delta S^{0,f}_{MgO(s)} = 2S^0_{MgO(s)} - 2S^0_{Mg(l)} - S^0_{O_2(g)}.$$

Because gas is a more random arrangement of atoms than liquid, $S^0_{Mg(g)} > S^0_{Mg(l)}$. Hence, $\Delta S^{0,f}_{MgO(s)}$ decreases at the boiling point of $Mg(l)$. Since from [4-27]

$$\left(\partial \Delta G^{0,f}_{MgO(s)} / \partial T\right)_p = -\Delta S^{0,f}_{MgO(s)},$$

the slope $-\Delta S^{0,f}_{MgO(s)}$ increases at 1105°C.

The reader should use similar arguments to explain the decrease in slope associated with the reaction involving the formation of PbO. Unlike the first reaction in which the *reactant* Mg changes phase, in the second reaction the *product* PbO changes phase.

Example Problem 4-6

Calculate the (a) entropy of fusion of Si at its normal melting point, and (b) Gibbs free energy of fusion of Si under unstable conditions at 1500°C. Assume $\Delta C_p = 0$.

Solution

(a) At the normal melting point, $\Delta G_{1693}^{f} = 0$. Substituting heat of fusion data from Table A.2 into [4-8],

$$\Delta G_{1693,\mathrm{Si}}^{f} = 0 = \Delta H_{1693,\mathrm{Si}}^{f} - (1693)\Delta S_{1693,\mathrm{Si}}^{f}$$
$$= 50{,}630 \text{ J/mol} - (1693)\Delta S_{1693,Si}^{f}$$

or

$$\Delta S_{1693,\mathrm{Si}}^{f} = \underline{29.91 \text{ J/(mol} \cdot \text{K)}}.$$

(b) Assuming $\Delta C_p = 0$, and substituting [2-19] and [3-17] into [4-8],

$$\Delta G_{1773,\mathrm{Si}}^{f} = \Delta H_{1693,\mathrm{Si}}^{f} - (1773)\Delta S_{1693,\mathrm{Si}}^{f}$$
$$= 50{,}630 \text{ J/mol} - (1773)(29.91 \text{ J/mol})$$
$$= \underline{-2400 \text{ J/mol}}.$$

Since $\Delta G_{1773,\mathrm{Si}}^{f} < 0$, melting is spontaneous above the normal melting point at 1 atm pressure.

Example Problem 4-7

Calculate the (a) entropy of transformation of $\mathrm{Ti}(\alpha)$ to $\mathrm{Ti}(\beta)$ at the normal transformation point of 882°C, and (b) Gibbs free energy of transformation of $\mathrm{Ti}(\alpha) \to \mathrm{Ti}(\beta)$ under unstable conditions at 930°C.

Solution

(a) At the normal transformation temperature, $\Delta G_{1155}^{\mathrm{Tr}} = 0$. Substituting transformation enthalpy data from Table A.2 into [4-8],

$$\Delta G_{1155,\mathrm{Ti}(s)}^{\mathrm{Tr}} = 0 = \Delta H_{1155,\mathrm{Ti}(s)}^{\mathrm{Tr}} - (1155)\Delta S_{1155,\mathrm{Ti}(s)}^{\mathrm{Tr}}$$
$$= 3473 - (1155)\Delta S_{1155,\mathrm{Ti}(s)}^{\mathrm{Tr}}$$

or $\quad \Delta S_{1155,\mathrm{Ti}(s)}^{\mathrm{Tr}} = \underline{3.0 \text{ J/(mol} \cdot \text{K)}}.$

(b) Substituting [2-16] and [3-16] into [4-8] and using heat capacity data from Table A.3A,

$$\Delta G_{1203,\mathrm{Ti}(s)}^{\mathrm{Tr}} = \Delta H_{1203,\mathrm{Ti}(s)}^{\mathrm{Tr}} - (1203)\Delta S_{1203,\mathrm{Ti}(s)}^{\mathrm{Tr}}$$
$$= \Delta H_{1155,\mathrm{Ti}(s)}^{\mathrm{Tr}} + \int_{1155}^{1203} \Delta C_p\, dT - (1203)\left[\Delta S_{1155,\mathrm{Ti}(s)}^{\mathrm{Tr}} + \int_{1155}^{1203} \Delta C_p\, dT/T\right].$$

·Computation of the final answer is left to the reader. The result is

$$\Delta G^{Tr}_{1203,\text{Ti}(s)} = \underline{-131.14 \text{ J/mol.}}$$

Since $\Delta G^{Tr}_{1203,\text{Ti}(s)} < 0$, the $\text{Ti}(\alpha) \to \text{Ti}(\beta)$ transformation is spontaneous above

the normal transformation temperature at 1 atm. Assuming $\Delta C_p = 0$, $\Delta G^{Tr}_{1203,\text{Ti}(s)}$
$= -136$ J/mol. The error is less than 4%.

4.4 TL ANALYSIS INCORPORATING PRESSURE DEPENDENCE

Solving for dH in [4-6] and substituting into [3-7]

$$dG + TdS + SdT = TdS + VdP$$

or

$$dG = VdP - SdT \tag{4-14}$$

At constant temperature [4-14] reduces to

$$dG = VdP \tag{4-15}$$

Integrating [4-15] assuming V is independent of P over the pressure interval,

$$\int_{G_1}^{G_2} dG = V \int_{P_1}^{P_2} dP$$

or

$$\Delta G = V \int_{P_1}^{P_2} dP \tag{4-16}$$

where

$$V = \frac{M}{\rho} = \frac{\text{molecular mass}}{\text{density}} = \left(\frac{\text{gm}}{\text{mol}} \cdot \frac{\text{cm}^3}{\text{gm}} \right)$$

$$= \text{molar volume } [\text{cm}^3/\text{mol}].$$

Example Problem 4-8

Calculate the minimum hydrostatic pressure required to stabilize Fe_3C at room temperature. Although Fe_3C is unstable at ordinary pressures, it exists as a separate phase in steel microstructures and is identified as a stable phase in the Fe-C phase diagram.

Solution

(1) **Set Up.** An isothermal TL is structured to include the formation of $Fe_3C(s)$ directly from its constituent elements at a constant temperature of 298 K and variable pressure. It is assumed that densities, hence volumes, remain constant over the pressure interval.

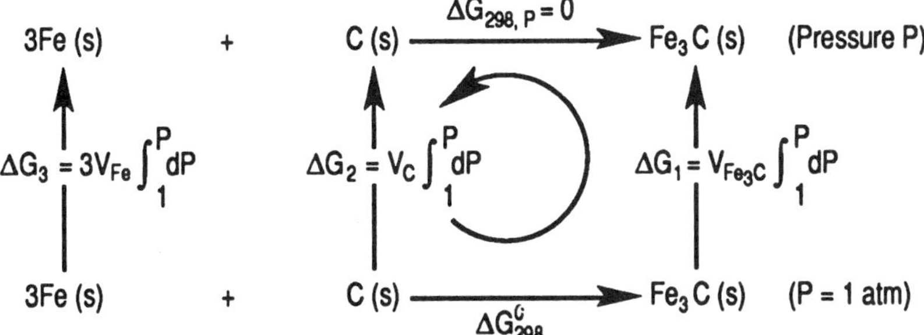

(2) Sum. Summing in the counterclockwise direction,

$$\sum \Delta G_{TL} = 0 = \Delta G_{298}^0 + \Delta G_1 - 0 - \Delta G_2 - \Delta G_3.$$

(3) Substitute. From Tables A.4 and B.1,

$$\Delta G_{298}^0 = 25{,}941 - 23.14(298) = 19{,}045 \text{ J/mol}$$
$$= (19{,}045 \text{ J/mol}) \times (82.057 \text{ cm}^3\text{·atm}/8.3144 \text{ J})$$
$$= 187{,}960 \text{ cm}^3\text{·atm/mol}.$$

$$\Delta G_1 = V_{Fe_3C} \int_1^P dP = (179.53 \text{ cm}^3/7.40 \text{ mol})(P - 1 \text{ atm})$$
$$= 24.24(P - 1) \text{ cm}^3\text{·atm/mol}.$$

$$\Delta G_2 = V_C \int_1^P dP = (12.01 \text{ cm}^3/2.25 \text{ mol})(P - 1 \text{ atm})$$
$$= 5.34(P - 1) \text{ cm}^3\text{·atm/mol}.$$

$$\Delta G_3 = 3V_{Fe} \int_1^P dP = 3(55.85 \text{ cm}^3/7.87 \text{ mol})(P - 1 \text{ atm})$$
$$= 21.29(P - 1) \text{ cm}^3\text{·atm/mol}.$$

Substituting into $\sum \Delta G_{TL} = 0$,
$$0 = 187{,}960 + 24.24(P - 1) - 0 - 5.34(P - 1) - 21.29(P - 1).$$

(4) Solve.

$$P \approx \underline{78{,}000 \text{ atm.}}$$

4.5 USEFUL THERMODYNAMIC RELATIONS

The concept of exactness as it pertains to thermodynamic properties has already been introduced in [1-4] and [1-5]. These equations and others are incorporated into relationships that are often found useful in solving a wide variety of thermodynamic problems in materials applications. A collection of these useful relationships is provided below.

(1) Condition of Exactness

If $z = f(x,y)$,

$$dz = \left(\frac{\partial z}{\partial x}\right)_y dx + \left(\frac{\partial z}{\partial y}\right)_x dy \qquad [1\text{-}4]$$

and

$$\left(\frac{\partial M}{\partial y}\right)_x = \left(\frac{\partial N}{\partial x}\right)_y \qquad [1\text{-}5]$$

where

$$M = M(x,y) = \left(\frac{\partial z}{\partial x}\right)_y \text{ and } N = N(x,y) = \left(\frac{\partial z}{\partial y}\right)_x .$$

(2) Introduction of a New Thermodynamic Variable, w

$$\left(\frac{\partial x}{\partial y}\right)_z = \left(\frac{\partial x}{\partial w}\right)_z \left(\frac{\partial w}{\partial y}\right)_z \qquad [4\text{-}17]$$

(3) Transformation Formula

Applying the definition of the total differential to $z = f(x,y)$, dividing by dy, and holding z constant:

$$\left(\frac{\partial z}{\partial y}\right)_z = 0 = \left(\frac{\partial z}{\partial x}\right)_y \left(\frac{\partial x}{\partial y}\right)_z + \left(\frac{\partial z}{\partial y}\right)_x \left(\frac{\partial y}{\partial y}\right)_z^{\;1}$$

or

$$\left(\frac{\partial x}{\partial y}\right)_z = -\frac{(\partial z / \partial y)_x}{(\partial z / \partial x)_y} \qquad [4\text{-}18]$$

Equation [4-18] is known as the *transformation formula*.

(4) Maxwell Equations

From Chapters 3 and 4,

$$dU = TdS - PdV \qquad [3\text{-}6]$$
$$dH = TdS + VdP \qquad [3\text{-}7]$$
$$dG = VdP - SdT \qquad [4\text{-}14]$$

Helmholtz free energy, A, is defined as

$$A = U - TS \qquad [4\text{-}19]$$

Taking the total derivative of [4-19],

$$dA = dU - TdS - SdT$$

and substituting for dU from [3-6],

$$dA = TdS - PdV - TdS - SdT$$

or

$$dA = -PdV - SdT \qquad [4\text{-}20]$$

For easy reference, the four *Maxwell equations* are:

$$dU = TdS - PdV \qquad [3\text{-}6]$$
$$dH = TdS + VdP \qquad [3\text{-}7]$$
$$dG = VdP - SdT \qquad [4\text{-}14]$$
$$dA = -PdV - SdT \qquad [4\text{-}20]$$

Applying [1-5], the Maxwell equations can be restated in the following alternate form:

$$\left(\frac{\partial T}{\partial V}\right)_S = -\left(\frac{\partial P}{\partial S}\right)_V \qquad [4\text{-}21]$$

$$\left(\frac{\partial T}{\partial P}\right)_S = \left(\frac{\partial V}{\partial S}\right)_P \qquad [4\text{-}22]$$

$$-\left(\frac{\partial S}{\partial P}\right)_T = \left(\frac{\partial V}{\partial T}\right)_P \qquad [4\text{-}23]$$

$$\left(\frac{\partial S}{\partial V}\right)_T = \left(\frac{\partial P}{\partial T}\right)_V \qquad [4\text{-}24]$$

Additionally, expressing [4-14] in partial derivative form, other useful relationships are:

$$\left(\frac{\partial G}{\partial T}\right)_P = -S \qquad [4\text{-}25]$$

$$\left(\frac{\partial G}{\partial P}\right)_T = V \qquad [4\text{-}26]$$

[4-25] and [4-26] may also be written in the form:

$$\left(\frac{\partial \Delta G}{\partial T}\right)_P = -\Delta S \qquad [4\text{-}27]$$

$$\left(\frac{\partial \Delta G}{\partial P}\right)_T = \Delta V \qquad [4\text{-}28]$$

(5) Gibbs-Helmholtz Equation

Substituting [4-27] into [4-8],

$$\Delta G = \Delta H + T\left(\frac{\partial \Delta G}{\partial T}\right)_P \qquad [4\text{-}29]$$

Dividing [4-29] by T^2 and rearranging,

$$\frac{1}{T}\left(\frac{\partial \Delta G}{\partial T}\right)_P - \frac{\Delta G}{T^2} = -\frac{\Delta H}{T^2}.$$

Applying the *differentiation theorem for a quotient of two functions*, ΔG and T (Protter and Morrey, 1970, p. 136),

$$\left[\frac{\partial(\Delta G / T)}{\partial T}\right]_P = \frac{1}{T}\left(\frac{\partial \Delta G}{\partial T}\right)_P - \frac{\Delta G}{T^2}.$$

Combining the last two equalities,

$$\left[\frac{\partial(\Delta G / T)}{\partial T}\right]_P = -\frac{\Delta H}{T^2} \qquad [4\text{-}30]$$

[4-30], known as the *Gibbs-Helmholtz equation*, is applicable to a *closed system* at constant pressure.

(6) Total Enthalpy

Applying the definition of the total differential to $H = f(T,P)$,

$$dH = \left(\frac{\partial H}{\partial T}\right)_P dT + \left(\frac{\partial H}{\partial P}\right)_T dP \qquad [4\text{-}31]$$

Dividing [3-7] by dP, holding temperature fixed, and assuming that molar volume is constant,

$$\left(\frac{\partial H}{\partial P}\right)_T = T\left(\frac{\partial S}{\partial P}\right)_T + V\left(\frac{\partial P}{\partial P}\right)_T$$

Substituting [4-23] and [1-2] into the above,

$$\left(\frac{\partial H}{\partial P}\right)_T = -T\left(\frac{\partial V}{\partial T}\right)_P + V$$

$$= -\alpha V T + V \qquad [4\text{-}32]$$

Substituting [4-32] and [2-12] into [4-31],

$$dH = C_p dT + V(1 - \alpha T)dP \qquad [4\text{-}33]$$

(7) Total Entropy

Applying the definition of the total differential to $S = f(T,P)$,

$$dS = \left(\frac{\partial S}{\partial T}\right)_P dT + \left(\frac{\partial S}{\partial P}\right)_T dP \qquad [4\text{-}34]$$

Substituting [4-23] and the differential form of [3-10] into [4-34],

$$dS = \frac{C_p \, dT}{T} - \left(\frac{\partial V}{\partial T}\right)_P dP \qquad [4\text{-}35]$$

Substituting [1-2] into [4-35],

$$dS = \frac{C_p \, dT}{T} - \alpha V \, dP \qquad [4\text{-}36]$$

4.6 DISCUSSION QUESTIONS

(4.1) Refer to Figure E.2. Give examples, where possible, of four different types of phase transformations for both elements and oxides.

(4.2) Consider element A in which the most stable standard state configuration at T_1 is α and at T_2 is β. From the definition of standard state Gibbs free energy of an element, answer the following and explain your reasoning:

(a) Is $\Delta G^0_{T_2,\alpha} = 0$?

(b) Is $\Delta G^0_{T_1,\beta} = 0$?

(c) Is $\Delta G^0_{T_1,\alpha} = 0$?

(d) Is $\Delta G^0_{T_2,\beta} = 0$?

(4.3) A phase transformation occurs reversibly at temperature T and pressure P. The phases involved are A and A$'$, and the transformations and corresponding Gibbs free energy changes are:

(1) $\text{A} \xrightarrow[(T,P)]{} \text{A}'$ $\qquad\qquad \Delta G^{Tr}_1$

(2) $\text{A}' \xrightarrow[(T,P)]{} \text{A}$ $\qquad\qquad \Delta G^{Tr}_2$

What is the relationship between ΔG^{Tr}_1 and ΔG^{Tr}_2?

(4.4) For which mineral reaction below is $\Delta G^0_T = \Delta G^{0,f}_{T,\text{Enstatite}}$? Explain.

(a) $SiO_2(s) + Mg_2SiO_4(s) \rightarrow 2MgSiO_3(s)$
   Quartz   Olivine   Enstatite

(b) $2Mg(s) + 2Si(s) + 3O_2(g) \rightarrow 2MgSiO_3(s)$

(4.5) Answer the following questions about Ellingham diagrams:

(a) How are ΔH^0 and ΔS^0 obtained from the diagram?

(b) How are the relative stabilities of oxides or sulfides compared directly from the diagram?

(c) Why does the slope of the line representing the oxidation of uranium decrease at $\approx 1300°C$?

(4.6) What effect does maintenance of a dynamic vacuum have on calculation of the standard Gibbs free energy change for a heterogenous reaction involving solids and gases?

(4.7) Is $\Delta H^0_{298,H(g)} = 0$? Explain.

4.7 EXERCISE PROBLEMS

[4.1] An ideal gas undergoes an isothermal process from an initial State 1: (P_1, V_1, T) to a final State 2: (P_2, V_2, T). Prove
$$\Delta G_T = RT \ln(P_2/P_1).$$

[4.2] The initial state of one mole of a monatomic ideal gas is $P_1 = 5$ atm and $T_1 = 300$ K. Calculate the entropy change of the gas for:

(a) an isothermal decrease in pressure to 1 atm.

Ans: $\Delta S = 13.38$ J/(mol·K).

(b) a constant volume decrease in the pressure to 1 atm.

Ans: $\Delta S = -20.07$ J/(mol·K).

[4.3] Prove by two methods that at constant temperature,
$$\Delta H_{\text{Isothermal}} = 0$$
for an ideal gas, regardless of the pressure path.

[4.4] Prove that at constant temperature,
$$\Delta S'_{\text{Isothermal}} = -nR \ln(P_2/P_1)$$
for n moles of an ideal gas.

[4.5] Calcium boils at 1440°C. The standard free energy of vaporization of liquid calcium is given by

$$\Delta G_T^0 \ (\text{cal/mol}) = 41{,}030 + 5.83T \log(T) - 42.23T,$$

where $\log = \log_{10}$ (personal communication, 1959, R. Schuhmann, Jr., Department of Metallurgical Engineering, Purdue University, West Lafayette, Indiana).

(a) Find the linear Gibbs free energy equation of the form
$$\Delta G_T^0 = a + bT$$
which approximates the above equation as closely as possible near 1500 K.

Ans: $\Delta G_T^0 \ (\text{cal/mol}) = 37{,}235 - 21.2T.$

(b) Calculate the boiling point from the linear equation and compare with the actual value.

Ans: $T = 1756$ K ($\approx 2.5\%$ error).

[4.6] Phosphorous-bearing vanadium deposits of the western U.S. are a source of vanadium used in the steel industry as an alloy component. X-ray examination of a representative sample of raw material from these deposits reveals that the major impurity present in the ore is iron in the form of Fe_2P. A suggested method for removing the iron is to convert Fe_2P to Fe_3P and pure iron (ferrite) by vacuum distillation of the phosphorous and to remove the iron magnetically. Assuming the best vacuum obtainable in a large scale operation is 10^{-2} mm Hg, estimate the operating temperature of a vacuum furnace required to accomplish this distillation.

Ans: $T \approx 1411$°C.

[4.7] Estimate the pressure which must be applied to increase the melting point of pure Au by 20°C.

Ans: P ≈ 3370 atm.

[4.8] Assuming that ΔH^0 and ΔS^0 are independent of temperature, calculate the temperature at which solid HgO will dissociate into liquid Hg and $O_2(g)$ at 1 atm.

Ans: $T \approx 567°C$.

[4.9] Thermodynamic analysis of the equilibrium state for carbon, chromium, and niobium carbide reveals that niobium is an effective alloying agent for limiting grain boundary precipitation of chromium carbide by preferentially removing carbon as niobium carbide.

(a) Write a single equation describing C-Nb-Cr equilibria neglecting solubility effects.

Ans: $6Nb + Cr_{23}C_6 \rightarrow 6NbC + 23Cr$.

(b) Confirm thermodynamically the alloying effect of Nb at 1200 K.

Ans: $\Delta G^0_{1200} = -311{,}200$ J/mol $< 0 \Rightarrow$ NbC formation is favored.

[4.10] Three equations for the oxidation of a metal M are given below. One of these equations is for the oxidation of solid M, one is for the oxidation of liquid M, and one is for the oxidation of gaseous M. Using the ΔG^0 data given, identify the reaction and the state of the reactant metal.

(a) $2M + O_2(g) \rightarrow 2MO(s)$ $\qquad \Delta G^0_T = -290{,}400 + 46.1T$

(b) $2M + O_2(g) \rightarrow 2MO(s)$ $\qquad \Delta G^0_T = -358{,}754 + 102.6T$

(c) $2M + O_2(g) \rightarrow 2MO(s)$ $\qquad \Delta G^0_T = -298{,}400 + 55.4T$

Ans: Reaction (a) solid M, (b) gaseous M; (c) liquid M.

[4.11] For the reaction $SiC(s) \rightarrow Si(s) + C(s)$, $\Delta G^0_T = 12{,}770 - 1.66T$ cal/mol from 298–1680 K. Using this data, determine:

(a) ΔS^0 at 1000 K.

Ans: $\Delta S^0_{1000} = 1.66$ cal/(mol·K).

(b) ΔH^0 at 1500 K.

Ans: $\Delta H^0_{1500} = 12{,}770$ cal/mol.

[4.12] Using the general expression for $C_p - C_v$ given in Exercise Problem [1.3]:

(a) Show that $C_p - C_v = \alpha^2 VT/\beta$ for any substance.

(b) Use this expression to estimate $C_p - C_v$ for pure copper at 298 K.

Ans: $C_p - C_v \approx 0.01$ J/(mol·K).

[4.13] Use the data in Exercise Problem [3.8] and the Gibbs free energy criteria to confirm that solidification is spontaneous after supercooling.

Ans: $\Delta G = -2877$ J/mol $< 0 \Rightarrow$ spontaneous.

[4.14] (a) Starting with [4-13], use [4-27] and [4-30] to compute ΔS^0 and ΔH^0, respectively, as a function of temperature.

Ans: $\Delta H_T^0 = A - \dfrac{BT}{2.303}$, $\Delta S_T^0 = \dfrac{B}{2.303}[\ln(1/T) - 1] - C$.

(b) Use the results from (a) and Appendix A, Table A.4 to obtain ΔH^0 and ΔS^0 as a function of temperature for the reaction

$$2Cu(s) + (1/2)O_2(g) \rightarrow Cu_2O(s).$$

Compare the results at $T = 298$ K with the data in Appendix A, Table A.1.

Ans: ΔH_{298}^0 (derived equation) = $-167{,}350$ J/mol,

ΔH_{298}^0 (Table A.1) = $-167{,}380$ J/mol,

ΔS_{298}^0 (derived equation) = -75.74 J/(mol·K);

ΔS_{298}^0 (Table A.1) = -75.27 J/(mol·K).

[4.15] The thermodynamics of SiO is important in many materials applications because of its high vapor pressure. Develop an expression for the Gibbs free energy of formation of SiO(l), $\Delta G_{SiO(l)}^{0,f}$, as a function of temperature between 298–1700 K using the data in Appendix A, Table A.4.

Ans: $\Delta G_{SiO(l)}^{0,f}$ (J/mol) = $-91{,}848 + 20.72T \log(T) - 150T$.

[4.16] One mole of a metal, m, at 1 atm pressure is heated at constant volume from 300 to 500 K. Calculate the hydrostatic pressure that results from such a process. What is the work done? Assume C_p is constant over the temperature interval. $\alpha_m = 5 \times 10^{-5}$ K^{-1}, and $\beta_m = 8 \times 10^{-6}$ atm^{-1}.

Ans: $P \approx 1251$ atm; $W = 0$ since V is constant.

[4.17] Referring to Exercise Problem [4.16], calculate the entropy change for the same process. Let $C_p^m = 20.9$ J/(mol·K), $V_m = 6$ cm^3/(gm mol).

Ans: $\Delta S_m = 10.64$ J/(mol·K).

[4.18] Use the Gibbs free energy criteria to confirm that liquid potassium and its vapor are in equilibrium at 950 K and 0.447 bar.

Ans: $\Delta g^{v,l}$ (kJ/kg) = $1979 - 1978 \approx 0 \Rightarrow$ equilibrium.

[4.19] Use [3-6] to confirm the value of $u_g - u_l$ in Appendix D, Table D.2, at 900 K and 0.251 bar.

Ans: $\Delta u = u_g - u_l = 1827$ kJ/kg.

[4.20] Referring to Exercise Problems [4.16] and [4.17], estimate the enthalpy change for the same process if $V_m = 10$ cm^3/mol. State assumptions in making the calculations.

Ans: $\Delta H \approx 5440$ J/mol, V_m and α_m are assumed constant.

[4.21] Using the standard enthalpy and entropy changes computed in Exercise Problems [2.10] and [3.14] for the reaction

$$Mg_3Si_2O_5(OH)_4(s) + 2SiO_2(s) \rightarrow Mg_3Si_4O_{10}(OH_2)(s) + H_2O(g),$$

calculate ΔG_T^0 at temperature T where 373 K $< T <$ 800 K. No phase changes occur.

Ans: ΔG_T^0 (J/mol) $= -82{,}315 - 4672.2T\ln(T) + 5.44 \times 10^3 T + 2.58T^2 + 3.47 \times 10^{-7}T^3 - 3.66 \times 10^7 T^{-1} - 2.63 \times 10^5 T^{0.5}$.

[4.22] Using the standard enthalpy and entropy changes computed in Exercise Problems [2.12] and [3.15] for the reaction

$$NaAlSi_3O_8(s) \rightarrow NaAlSi_2O_6(s) + SiO_2(s),$$

calculate ΔG_T^0 at temperature T where 298 K $< T <$ 844 K. No phase changes occur.

Ans: ΔG_T^0 (J/mol) $= 53.21 \times 10^3 - 2000.4T + 238.21T\ln(T) - 70.349 \times 10^{-3}T^2 + 3.791 \times 10^{-6}T^3 + 17.478 \times 10^3 T^{0.5} + 2460.4 \times 10^3 T^{-1}$.

[4.23] The sorosilicate lawsonite, $CaAl_2Si_2O_7(OH)_2 \cdot H_2O(s)$, is a common constituent of metamorphic rocks formed at low temperature and high pressure. Assuming the molar volume of this mineral is independent of pressure, calculate the molar isothermal Gibbs free energy change for lawsonite subjected to a reversibly applied pressure increase from 1 to 5 bar at 298 K.

Ans: $\Delta G = 40.5$ J/mol.

[4.24] Using the standard enthalpy and entropy of formation computed in Exercise Problems [2.13] and [3.13] for the reaction

$$2Cu(l) + 1/2S_2(g) \rightarrow Cu_2S(\gamma),$$

determine whether or not this reaction is spontaneous at 1356 K as written.

Ans: $\Delta G_{1356}^0 = -274{,}260$ J/mol $< 0 \Rightarrow$ spontaneous.

[4.25] For the contact metamorphic reaction

$$SiO_2(\alpha) + CaCO_3(s) \rightarrow CaSiO_3(s) + CO_2(g),$$

(a) Calculate the standard Gibbs free energy change at temperature 298 K $\leq T <$ 844 K. Assume $\Delta C_p = 0$ and use the value of ΔS_{298}^0 computed in Example Problem [3.12].

Ans: ΔG_T^0 (J/mol) $= 88{,}670 - 165.65T$.

(b) Plot ΔG_T^0 as a function of T. Label line segments and points corresponding to a reversible reaction, a spontaneous reaction, and a nonspontaneous reaction as written.

Ans: $T^{Rev} = 535$ K, $T^{Irr} > 535$ K; $T^{Nonspon.} < 535$ K.

[4.26] Show that the Gibbs free energy change for an isobaric chemical reaction occurring at a temperature T is

$$\Delta G_T = \Delta H_{T_1} - T\Delta S_{T_1}$$

when $\Delta C_p = 0$ in the temperature range between T_1 to T.

[4.27] Derive [4-13] showing all steps.

[4.28] Derive [4-14] and show that

$$-\left(\frac{\partial S}{\partial P}\right)_T = \left(\frac{\partial V}{\partial T}\right)_P.$$

HETEROGENEOUS EQUILIBRIA: VARIABLE GAS PHASE COMPOSITION

5.1 IDEAL GAS MIXTURES, DALTON'S LAW, AND GIBBS FREE ENERGY OF A GAS

In most cases of interest in this book, gases are at low enough pressure to exhibit ideal behavior as expressed by the ideal gas law, [1-1]. Additionally, gases at these pressures are normally treated as *ideal mixtures* (Chapter 6, Section 6.3). *Dalton's law* states that the total pressure P_T of an *ideal gas mixture* is equal to the sum of the pressures exerted by each component. For example, the *partial pressure*, P_i, is the pressure that component i alone would exert if it occupied the same volume as the gas mixture at the same temperature. By Dalton's law,

$$P_T = \Sigma P_i \qquad [5\text{-}1]$$

Since each pure gas component occupies a total mixture volume of V' at temperature T, P_i can be expressed by rewriting [1-1] as

$$P_i = \frac{n_i RT}{V'} \qquad [5\text{-}2]$$

Substituting [5-2] into [5-1],

$$P_T = \left(\Sigma n_i\right)\frac{RT}{V'} \qquad [5\text{-}3]$$

Dividing [5-2] by [5-3],

$$\frac{P_i}{P_T} = \frac{n_i}{\Sigma n_i} = Y_i$$

or

$$P_i = Y_i P_T \qquad [5\text{-}4]$$

where Y_i is the mole fraction of component i in the ideal gas mixture. [5-4], a useful alternate form of Dalton's law, states that the partial pressure of component i in an ideal gas mixture can be obtained by multiplying the mole fraction of that component by the total gas pressure. Substituting the molar volume V of an ideal gas component i in a mixture from [1-1] into [4-14] at constant temperature and integrating,

$$G_i - G_i^0 = \Delta G_i = \int_{P_i^0}^{P_i} \frac{RT}{P_i}dP_i$$

$$= RT\ln(P_i / P_i^0) \qquad [5\text{-}5]$$

The *standard state pressure*, P_i^o, of an ideal gas, insoluble in any condensed phase with which it is in contact, is chosen to be 1 atm (Chapter 6, Section 6.2) at *any temperature*. Inserting [5-4] into [5-5] at $P_i^0 = 1$ atm,

$$\Delta G_i = RT \ln(Y_i P_T) \qquad [5\text{-}6]$$

From [5-4], $$\Delta G_i = RT \ln(P_i) \qquad [5\text{-}7]$$

5.2 TL ANALYSIS AND ELLINGHAM DIAGRAMS

TL analysis of *heterogeneous equilibria** is demonstrated below with a variety of Example Problems. In this chapter, all condensed phases are pure, hence, $\Delta G_T = 0$. This result follows from [4-14] since at constant or relatively low pressures, $VdP \approx 0$ and at constant temperature, $dT = 0$.

Example Problem 5-1

Calculate the equilibrium P_{O_2} over Ni at 1200°C. Determine the corresponding vacuum below which NiO will begin to dissociate. Give answers in mm Hg.

Solution

(1) Set Up.

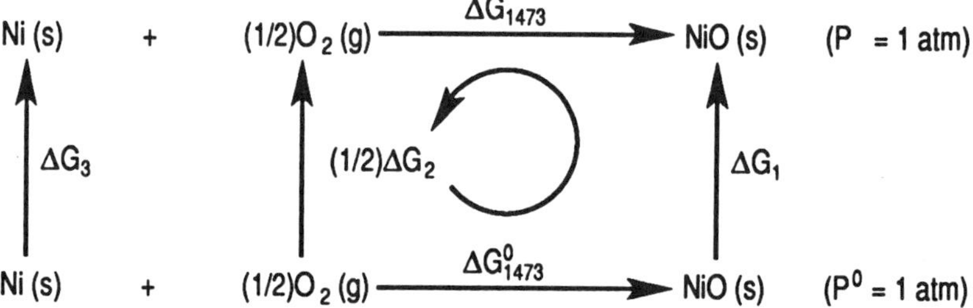

(2) Sum. Summing in the counterclockwise direction,

$$\Sigma\Delta G_{TL} = 0 = \Delta G_{1473}^0 + \Delta G_1 - \Delta G_{1473} - (1/2)\Delta G_2 - \Delta G_3.$$

(3) Substitute. From Table A.4,

$$\Delta G_{1473}^0 = -244{,}580 + 98.54(1473) = -99{,}431 \text{ J/mol.}$$

$\Delta G_{1473} = 0$ at equilibrium.

$\Delta G_1 = \Delta G_3 = 0$ for pure condensed phases.

$$(1/2)\Delta G_2 = (1/2)RT \ln\left(P_{O_2}\right) = 8.3144(1473)\ln\left(P_{O_2}^{1/2}\right).$$

---

* Heterogeneous equilibria refers to component equilibrium in systems of two or more phases.

Substituting into $\Sigma\Delta G_{TL} = 0$,
$$0 = -99{,}431 - 8.3144 \times 1473\ln\left(P_{O_2}^{1/2}\right).$$

(4) Solve.
$$\ln\left(P_{O_2}^{1/2}\right) = -99{,}431/8.3144(1473), \text{ hence}$$
$$P_{O_2} = 8.9 \times 10^{-8} \text{ atm} \times (760 \text{ mm Hg/atm})$$
$$= \underline{6.8 \times 10^{-5} \text{ mm Hg}}.$$

The vacuum below which the reaction will shift from oxidation to deoxidation is found from Dalton's law. In air, $Y_{O_2} \approx 0.22$ ($Y_{N_2} \approx 0.78$). Substituting P_{O_2} and Y_{O_2} into [5-4],
$$P_T = 6.8 \times 10^{-5} \text{ mm Hg}/0.22 = \underline{3.1 \times 10^{-4} \text{ mm Hg}}.$$

The above results can be confirmed from the Ellingham diagram, Figure E.2. A nomograph, superimposed on this diagram, is used to determine the equilibrium P_{O_2} directly. Figure 5.1, a simplified portion of Figure E.2, illustrates the graphical procedure. A straight line is drawn from the point labeled O (top of the vertical line on the left side of the diagram) so as to intersect the NiO reaction line at 1200°C. This line is extrapolated to the right hand scale corresponding to the P_{O_2}. Interpolating,
$$P_{O_2} \approx 5 \times 10^{-8} \text{ atm} \times (760 \text{ mm Hg/atm}) = 3.8 \times 10^{-5} \text{ mm Hg}.$$

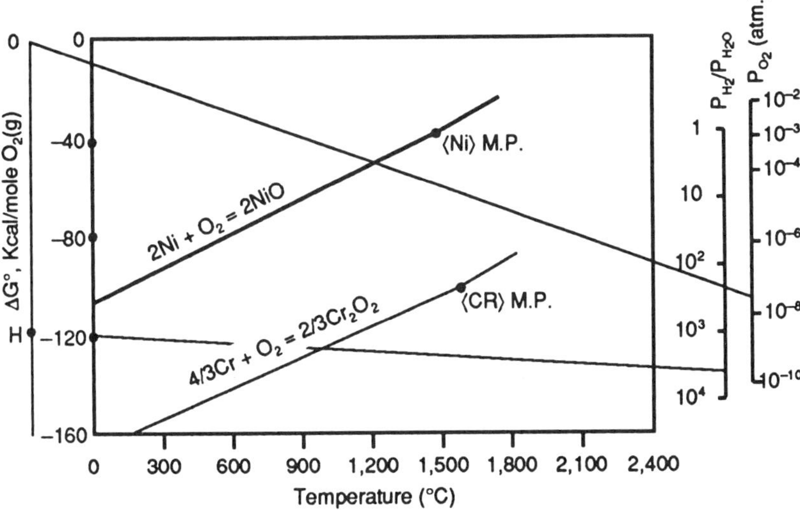

Figure 5.1 Nomographic determination of equilibrium gas partial pressures over metallic Ni at 1200°C and Cr at 1000°C. Points O and H on the vertical line at the left side of the Ellingham diagram are extended through the selected temperature/reaction point to the vertical scale at the right as illustrated.

Substituting into [5-4], $P_T = P_{O_2}/Y_{O_2} = 3.8 \times 10^{-5}$ mm Hg/0.22 $= 1.7 \times 10^{-4}$ mm Hg. Extrapolation from the diagram is somewhat less precise than the technique utilizing TL analysis because of graphical and interpolation errors.

In a manner analogous to that just presented, Ellingham diagrams can also be used to graphically estimate the equilibrium ratios P_{H_2}/P_{H_2O} and P_{CO}/P_{CO_2} in oxide equilibria. The next example demonstrates the procedure after first presenting the solution using TL analysis.

Example Problem 5-2

Calculate the equilibrium ratio P_{H_2}/P_{H_2O} for the oxidation of chromium in water vapor at 1000°C.

Solution

(1) Set Up.

$$(4/3)Cr(s) \quad + \quad 2H_2O(g) \xrightarrow{\;\Delta G_{1273}\;} (2/3)Cr_2O_3(s) \quad + \quad 2H_2(g)$$

$$(4/3)\Delta G_4 \qquad 2\Delta G_3 \qquad\qquad (2/3)\Delta G_2 \qquad 2\Delta G_1$$

$$(4/3)Cr(s) \quad + \quad 2H_2O(g) \xrightarrow{\;\Delta G^0_{1273}\;} (2/3)Cr_2O_3(s) \quad + \quad 2H_2(g)$$

(2) Sum.

$$\Sigma\Delta G_{TL} = 0 = \Delta G^0_{1273} + 2\Delta G_1 + (2/3)\Delta G_2 - \Delta G_{1273} - 2\Delta G_3 - (4/3)\Delta G_4.$$

(3) Substitute. From Table A.4,

$$\Delta G^0_{1273} = (2/3)[-1,120,370 + 259.85(1273)] + 2[-246,460 + 54.82(1273)]$$
$$= -173,039 \text{ J.}$$

$\Delta G_{1273} = 0$ at equilibrium.

$2\Delta G_1 = 2R(1273) \ln\left(P_{H_2}\right)$ and $2\Delta G_3 = 2R(1273) \ln\left(P_{H_2O}\right)$.

$(2/3)\Delta G_2 = (4/3)\Delta G_4 = 0.$

Substituting the above into $\Sigma\Delta G_{TL} = 0$,

$$0 = -173,039 + 2(8.3144)(1273) \ln\left(P_{H_2}\right) - 2(8.3144)(1273)\ln(P_{H_2O}).$$

(4) Solve.

$$P_{H_2}/P_{H_2O} = \underline{3.55 \times 10^3}.$$

Note that if $P_{H_2}/P_{H_2O} > 3.55 \times 10^3$, $Cr_2O_3(s)$ will tend to be reduced because excess P_{H_2} drives the reaction to the left. If $P_{H_2}/P_{H_2O} < 3.55 \times 10^3$, $Cr(s)$ will tend to oxidize because excess P_{H_2O} drives the reaction to the right. A qualitative statement of this behavior, known as *LeChatelier's principle*, states that "when a system, which is at equilibrium, is subjected to the effects of an external influence, the system moves in that direction which tends to nullify the effects of the external influence" (Gaskell, 1981, p. 132).

Using the nomograph on Figure 5.1, a straight line is drawn from the point labeled H so as to intersect the Cr_2O_3 reaction line at 1000°C. Extrapolation to the right hand scale gives $P_{H_2}/P_{H_2O} \approx 8 \times 10^3$. The procedure for graphically determining the P_{CO}/P_{CO_2} ratio for the oxidation of an oxide in CO_2 gas is analogous to that used in the last two examples except that point C (Figure E.2, left hand vertical line) is used in conjunction with the P_{CO}/P_{CO_2} scale.

Example Problem 5-3

Characterize the relative thermal stability of Si_3N_4 and BN in a mixture of Si_3N_4 and B at 1 atm pressure. Assume all components are pure. BN has been suggested as an abradable high temperature coating. Si_3N_4 is utilized in high temperature ceramic applications.

Solution

(1) Set Up.

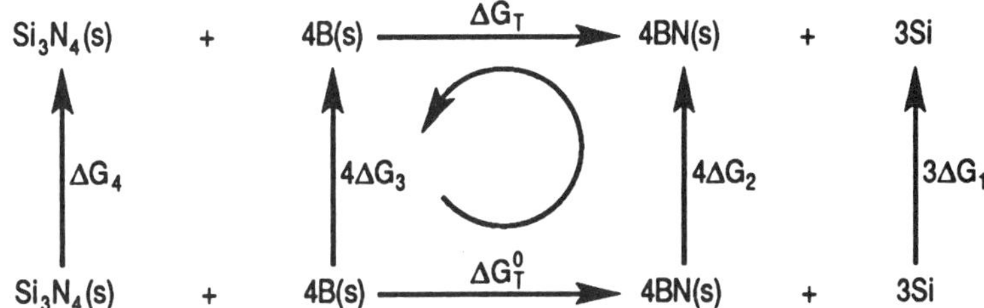

(2) Sum.

$$\Sigma \Delta G_{TL} = 0 = \Delta G_T^0 + 3\Delta G_1 + 4\Delta G_2 - \Delta G_T - 4\Delta G_3 - \Delta G_4$$

or

$$\Delta G_T = \Delta G_T^0 + 3\Delta G_1 + 4\Delta G_2 - 4\Delta G_3 - \Delta G_4.$$

(3) Substitute.

$\Delta G_T = 0$ at equilibrium.

$3\Delta G_1 = 4\Delta G_2 = 4\Delta G_3 = \Delta G_4 = 0.$

$\Delta G_T^0 = 0.$

The problem now becomes one of finding ΔG_T^0 as a function of temperature and setting that expression equal to zero. The value of ΔG_T^0 depends on whether Si is in the solid or liquid state, for which

$$\Delta G_T^0 = 4\Delta G_{T,BN(s)}^0 - \Delta G_{T,Si_3N_4(s)}^0$$

from [4-9]. Note that the f for Gibbs free energy of formation has been dropped. From Table A.4:

Assuming Si is solid,

$$2B(s) + N_2(g) = 2BN(s) \qquad \Delta G_T^0 = -217{,}590 + 81.18T \text{ J/mol}$$
from 1200-2300 K

$$2N_2(g) + 3Si(s) = Si_3N_4(s) \qquad \Delta G_T^0 = -753{,}190 + 336.43T \text{ J/mol}$$
from 298-1680 K

Combining Gibbs free energies, substituting the result into [4-9], and setting the expression equal to zero:

$$0 = 4[(-217{,}590 + 81.18T)/2] - (-753{,}190 + 336.43T).$$

(4) **Solve.** Solving the above expression, $\underline{T = 1827 \text{ K}}$. Since $T = 1827$ K is outside the allowable temperature range for $\Delta G_{Si_3N_4(s)}^0$, the calculation is *invalid*.

Assuming Si is liquid,

$$2B(s) + N_2(g) = 2BN(s) \qquad \Delta G_T^0 = -217{,}590 + 81.18T \text{ J/mol}$$
from 1200–2300 K

$$2N_2(g) + 3Si(l) = Si_3N_4(s) \qquad \Delta G_T^0 = -892{,}950 + 419.28T \text{ J/mol}$$
from 1680–1800 K

Combining Gibbs free energies, substituting the result into [4-9], and setting the expression equal to zero:

$$0 = 4[(-217{,}590 + 81.18T)/2] - (-892{,}950 + 419.28T).$$

Solving, $\underline{T = 1782 \text{ K}}$. Since T = 1782 K is within the allowable temperature range for $Si_3N_4(s)$, the calculation is *valid*. It can be concluded that BN(s) is stable above 1782 K whereas $Si_3N_4(s)$ is stable below 1782 K. Depending upon reaction kinetics, BN(s) as a coating on $Si_3N_4(s)$ would be satisfactory above $\approx$1782 K.

A graphical solution to the previous problem is obtained from a simplified portion of Figure E.4 shown in Figure 5.2. The intersection of the BN(s) and $Si_3N_4(s)$ lines yields an equilibrium temperature of $\approx$1500°C or 1773 K. Agreement with the computed result is excellent. Stability fields for these two compounds are identified. The field to the left of the intersection is stable BN(s). The field to the right is stable $Si_3N_4(s)$.

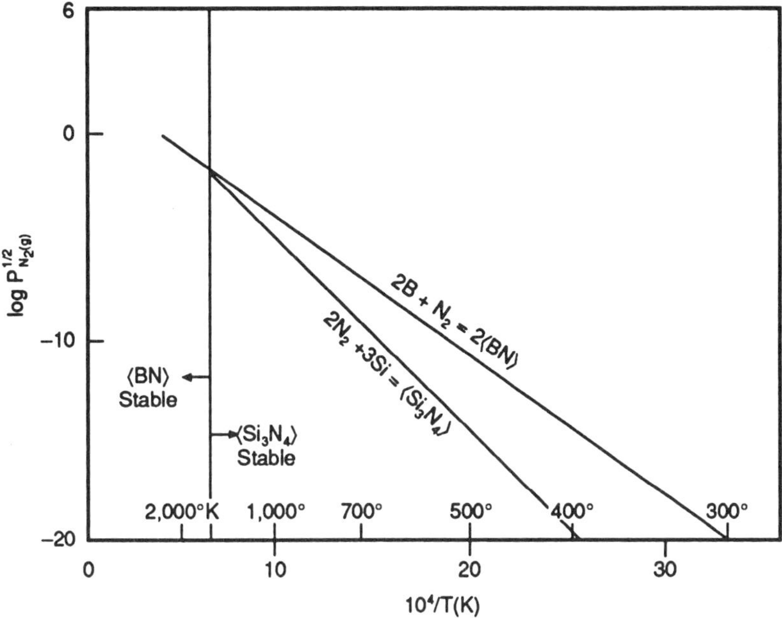

Figure 5.2 Log $\left(P_{N_2}^{1/2}\right)$ vs. $10^4/T$ for BN(s) and Si$_3$N$_4$(s) equilibria.

Example Problem 5-4

Calculate the equilibrium ratio P_{H_2S}/P_{H_2} for the formation of MnS at 1500°C. The presence of MnS inclusions in steel can be critical to fracture toughness in hydrogen environments.

Solution

(1) Set Up.

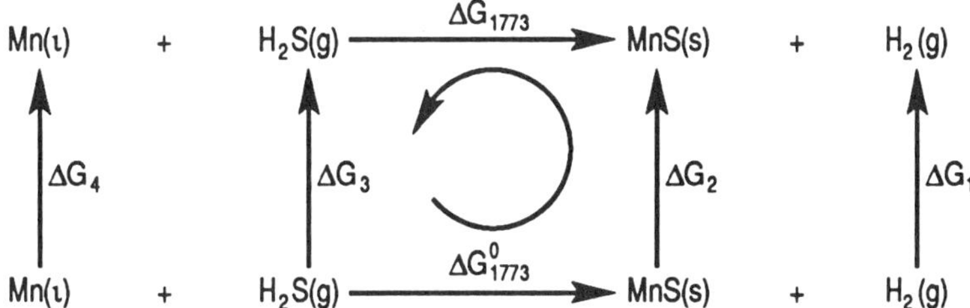

(2) Sum.

$$\Sigma \Delta G_{TL} = 0 = \Delta G_{1773}^0 + \Delta G_1 + \Delta G_2 - \Delta G_{1773} - \Delta G_3 - \Delta G_4.$$

(3)   Substitute.

$\Delta G_2 = \Delta G_4 = \Delta G_{1773} = 0.$

$\Delta G_1 = R(1773)\ln(P_{H_2})$ and $\Delta G_3 = R(1773)\ln(P_{H2S}).$

$\Delta G^0_{1773} = \Delta G^0_{1773,MnS(s)} - \Delta G^0_{1773,H_2S(g)}.$

Using the data in Table A.4 to determine ΔG^0_{1773},

$\Delta G^0_T = -198{,}470 + 29.53T$ J/mol.

$\Delta G^0_{1773} = -198{,}470 + 29.53(1773) = -146{,}113$ J/mol.

Substituting above into $\Sigma\Delta G_{TL} = 0$,

$0 = -146{,}113 + R(1773)[\ln(P_{H_2}) - \ln(P_{H_2S})].$

(4)   Solve.

$$\ln\left(\frac{P_{H_2}}{P_{H_2S}}\right) = \frac{146{,}113 \text{ J/mol}}{[8.3144 \text{ J/(mol}\cdot\text{K})](1773\text{K})} \Rightarrow \frac{P_{H_2}}{P_{H_2S}} = 20{,}165$$

or

$$P_{H_2S}/P_{H_2} \approx \underline{5\times10^{-5}}.$$

Figure 5.3, a simplified version of Figure E.3, offers an alternate solution. Using the nomograph, $P_{H_2S}/P_{H_2} \approx 8\times10^{-5}$.

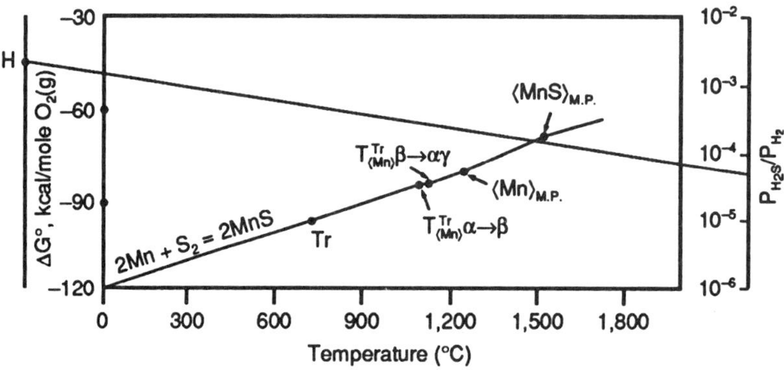

Figure 5.3 Nomographic determination of P_{H_2S}/P_{H_2} over Mn at 1500°C.

Example Problem 5-5

Li$_3$N formation has been thought to interrupt bonding between silver base-lithium alloys and titanium during brazing. Determine the temperature at which tri-lithium nitride begins to dissociate. Use Figure 5.4 (a portion of Figure E.4) to develop an expression for $\Delta G^0_{T,Li_3N(s)}$ (also see Exercise Problems [5.5] and [5.10]).

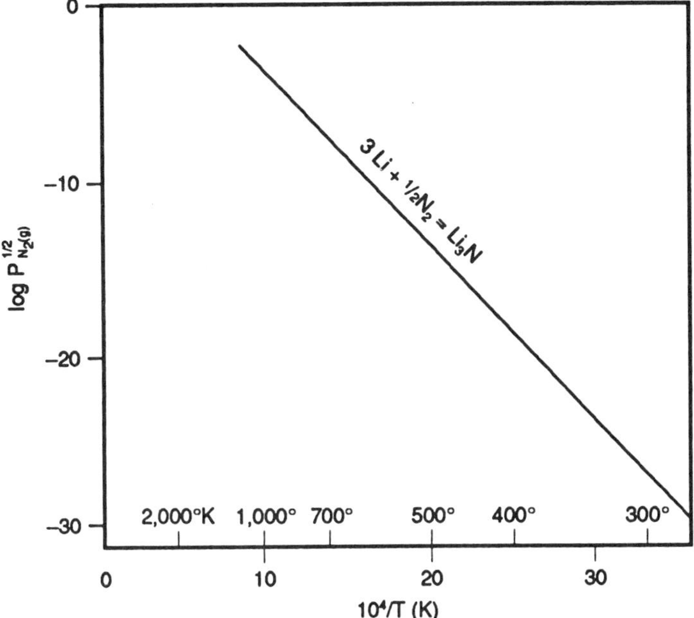

Figure 5.4 $\mathrm{Log}\left(P_{N_2}^{1/2}\right)$ vs. $10^4/T$ for lithium-nitrogen equilibria.

Solution

(1) Set Up.

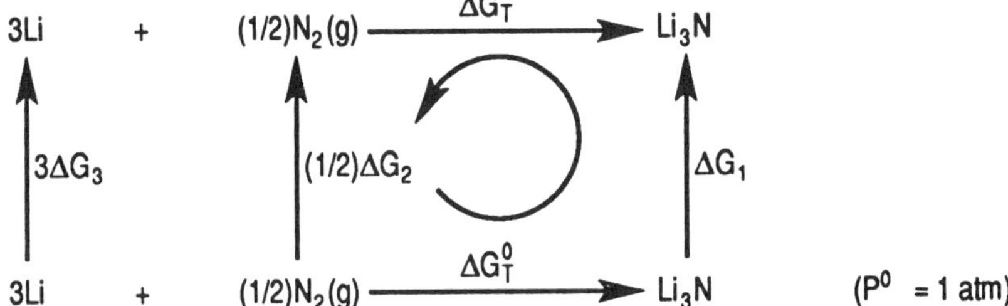

(2) Sum.

$$\Sigma \Delta G_{TL} = 0 = \Delta G_T^0 + \Delta G_1 - \Delta G_T - (1/2)\Delta G_2 - 3\Delta G_3.$$

(3) Substitute.

$$\Delta G_1 = 3\Delta G_3 = 0.$$

$$(1/2)\Delta G_2 = (1/2)RT\ln\left(P_{N_2}\right) = RT\ln\left(P_{N_2}^{1/2}\right).$$

Substituting into $\Sigma \Delta G_{TL} = 0$,

$$0 = \Delta G_T^0 - \Delta G_T - RT\ln\left(P_{N_2}^{1/2}\right).$$

(4)   Solve.

$$\Delta G_T^0 = \Delta G_T + RT \ln\left(P_{N_2}^{1/2}\right) = \Delta G_T + RT(2.303) \log\left(P_{N_2}^{1/2}\right).$$

ΔG_T is unknown except at equilibrium for which $\Delta G_T = 0$. Assuming $\Delta C_p \approx 0$, [4-10] may be used with equilibrium partial pressure data from Figure 5.4 to write two equations. Simultaneous solution will give ΔG_T^0. Selecting $\log\left(P_{N_2}^{1/2}\right)$ values at two temperatures and substituting into the above expression:

$$\Delta G_{453\,K}^0 = (8.3144)(453)(2.303)(-15) = \Delta H_T^0 - (453)\,\Delta S_T^0;$$

$$\Delta G_{298\,K}^0 = (8.3144)(298)(2.303)(-27) = \Delta H_T^0 - (298)\,\Delta S_T^0$$

where $\log\left(P_{N_2}^{1/2}\right) \approx -15$ at 453 K and $\log\left(P_{N_2}^{1/2}\right) = -27$ at 298 K. Solving the two Gibbs free energy equations simultaneously,

$$\Delta H_T^0 = -200,100;$$
$$\Delta S_T^0 = -154.5.$$

The desired expression is

$$\Delta G_T^0 = \underline{-200,100 + 154.5T}\ \text{J/mol}.$$

Setting $\Delta G_T^0 = 0$, T $\approx$ <u>1295 K or 1022°C</u> (Note: this result is approximate since the expression for ΔG_T^0 is strictly valid for 250 K $< T <$ 453 K).

Example Problem 5-6

Characterize the thermodynamics of the epitaxial formation of pure solid Si on a metallic substrate according to the chemical reaction:

$$SiCl_4(g) + 2H_2(g) \xrightarrow{\ 1200°C\ } Si(s) + 4HCl(g).$$

Note that both reactants and HCl are gases, while Si remains solid at 1200°C. The Si(s) product is a high purity, thin single crystal layer. $P_T = 1$ atm. An understanding of thermodynamics is an important step in understanding coating processes.

Solution

(1)   Set Up.

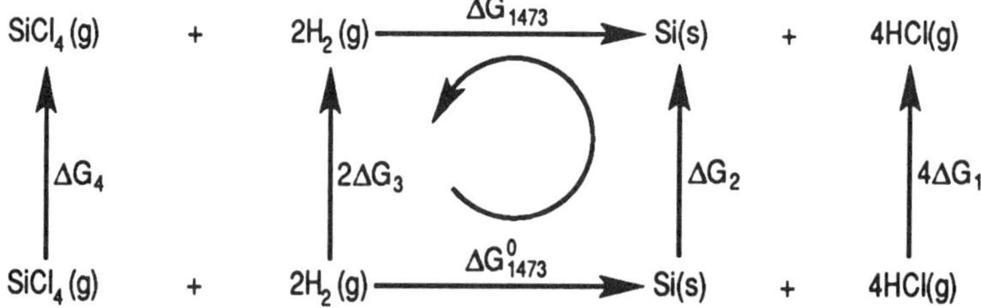

(2)   Sum.

$$\Sigma\Delta G_{TL} = 0 = \Delta G_{1473}^0 + 4\Delta G_1 + \Delta G_2 - \Delta G_{1473} - 2\Delta G_3 - \Delta G_4.$$

(3) Substitute. From Figure E.5,

$$\Delta G^0_{1473} = 4\Delta G^0_{HCl(g)} - \Delta G^0_{SiCl_4(g)} = -99{,}000 - (-98{,}000) \text{ cal/mol}$$

$$= -4183 \text{ J/mol Si.}$$

$$4\Delta G_1 = RT \ln\left(P^4_{HCl}\right).$$

$$\Delta G_2 = \Delta G_{1473} = 0.$$

$$2\Delta G_3 = RT \ln\left(P^2_{H_2}\right).$$

$$\Delta G_4 = RT \ln\left(P_{SiCl_4}\right).$$

Substituting into $\Sigma\Delta G_{TL} = 0$,

$$0 = -4183 + RT \ln\left(P^4_{HCl}\right) - RT \ln\left(P^2_{H_2}\right) - RT \ln\left(P_{SiCl_4}\right).$$

(4) Solve.

$$\ln\left(\frac{P^4_{HCl}}{P^2_{H_2} \cdot P_{SiCl_4}}\right) = \frac{4183 \text{ J/mol}}{[8.3144 \text{ J/(mol}\cdot\text{K)}](1473 \text{ K})} = 0.342$$

or

$$\frac{P^4_{HCl}}{P^2_{H_2} \cdot P_{SiCl_4}} \approx 1.4.$$

Since there are three gas components, three independent expressions incorporating 3 gases are required to determine the partial pressure of each gas. Note that Si remains solid and does not enter into the calculations.

(1) Assume, from the stoichiometry of the reaction, that 1 mole of $SiCl_4(g)$ and 2 moles of $H_2(g)$ are *initially* available to react.

(2) Let n = the number of moles of $SiCl_4(g)$ that reacts with $2n$ moles of $H_2(g)$ to form $Si(s)$ and $HCl(g)$.

(3)
Gas	$SiCl_4(g)$	$H_2(g)$	$HCl(g)$
Initial Moles	1	2	0
Moles Reacting or Produced	n	$2n$	$4n$
Moles Remaining	$1-n$	$2-2n$	$4n$

The total number of moles as a function of n are $(1-n) + (2-2n) + 4n = 3 + n$.

(4) Applying [5-4],

$$P_{SiCl_4(g)} = \frac{1-n}{3+n} \qquad P_{H_2(g)} = \frac{2-2n}{3+n} \qquad P_{HCl(g)} = \frac{4n}{3+n}$$

where partial pressure is in atmospheres and $P_T = 1$ atm.
Substituting into the partial pressure expression at 1200°C,

$$\frac{\left(\dfrac{4n}{3+n}\right)^4}{[(2-2n)/(3+n)]^2(1-n)/(3+n)}=1.4\,.$$

Expanding and solving for n,

$$65.4n^4 - 8.4n^2 + 11.2n - 4.2 = 0 \Rightarrow$$
$$n \approx 0.368.$$

Substituting into each gas partial pressure expression:

$$P_{SiCl_4(g)} = (1 - 0.368)/3.368 = \underline{0.188\ atm},$$

$$P_{H_2(g)} = [2 - 2(0.368)]/3.368 = \underline{0.375\ atm},$$

$$P_{HCl(g)} = 4(0.368)/3.368 = \underline{0.437\ atm};$$

$$P_{Total} = 1\ atm\ (check).$$

5.3 DISCUSSION QUESTIONS

(5.1) An ideal gas mixture contains gas component i. In accordance with Dalton's law:

    (a) Is the total volume of i the same as the total volume of the gas mixture? Why?

    (b) Is the molar volume of i the same as the molar volume of the gas mixture? Why?

(5.2) Assume the reaction A$\rightarrow$ B is at equilibrium. Is $\Delta G_T^0 = 0$?

(5.3) Are $H_2O(g)$ and $H_2O(l)$ polymorphs?

(5.4) How does the partial pressure of a gas component in an ideal gas mixture vary with respect to the mole fraction of that component?

(5.5) Does the Gibbs free energy of a gas in an isothermal ideal gas mixture increase or decrease when the partial pressure of the gas increases? Explain.

5.4 EXERCISE PROBLEMS

[5.1] An ideal gas mixture at pressure P_{Total} contains a gas component i of atomic fraction Y_i. Show
$$G_i - G_i^0 = RT[\ln(P_{Total}) + \ln(Y_i)].$$

[5.2] An isothermal ideal gas mixture changes pressure from $P_{1,Total}$ to $P_{2,Total}$. The mole fraction of gas i in the mixture is Y_i. Prove
$$\Delta G_i = RT \ln(P_{2,Total}/P_{1,Total}).$$

[5.3] During carburization processing of steel at 870°C in a mixture of CO and CO_2, surface oxidation of iron may prevent penetration of carbon if P_{CO_2} is too high. Calculate the equilibrium P_{CO}/P_{CO_2} ratio. What effect does a higher than equilibrium ratio have on oxidation? Use Appendix A, Table A.4 data. Compare the results obtained with Figure E.2 in Appendix E.

Ans: P_{CO}/P_{CO_2} = 1.78 (compared to ≈ 2.0 from Figure E.2). A higher than equilibrium ratio is deoxidizing.

[5.4] It is difficult to prevent the formation of MnS during the steel making process. One possible technique in reducing MnS content is to *getter* or selectively remove S by adding an element to the liquid steel bath that forms a more stable sulfide than manganese. Use Figure E.3 in Appendix E to select a possible addition. Identify the parameters used in making the selection and state assumptions.

Ans: Figure E.3 reveals that Ca may be a good selection since S reacts preferentially with Ca. The process is defined as *gettering*.

[5.5] The Gibbs free energy change for the reaction given in Example Problem 5-5 is provided by Wicks and Block (1963, p. 68):
ΔG_T^0 (cal/mol) = $- 47,000 + 12.4T \ln(T) - 11.36 \times 10^{-3}T^2 + 0.54 \times 10^4 T^{-1} - 37.8T$ where 453 K $\leq T \leq$ 800 K. Using this information,

(a) Calculate ΔG^0 at 800 K.

Ans: ΔG_{800}^0 = –76,120 J/mol.

(b) Compare the answer from (a) with the answer calculated from Figure 5.4. Discuss the difference. Also see Exercise Problem [5.10].

Ans: ΔG_{800}^0 = –76,500 J/mol. The comparison is good considering the extrapolation of the expression in Example Problem 5-5.

[5.6] It is proposed that a nickel-based alloy (assume pure Ni for this problem) be used in a structural application involving contact with SO_2.

(a) Find the partial pressure of S_2 and SO_2 in the contact gas when an equilibrium mixture is brought together at 812°C and a total pressure of 1 atm.

Ans: P_{S_2} = 0.12 atm; P_{SO_2} = 0.88 atm.

(b) Derive an expression relating the partial pressure of S_2 (P_{S_2}) as a function of total pressure (P_T). Calculate P_{S_2} and P_{SO_2} when P_T = 2 atm.

Ans: $P_{S_2} = P_T + 3.305 \pm 2.57\sqrt{P_T + 1.652}$, P_{S_2} = 0.39 atm; P_{SO_2} = 1.61 atm.

(c) What is the significance of the increase in total pressure on oxidation?

Ans: As P_{Total} increases, so does the tolerable P_{SO_2} before appreciable Ni oxidation occurs.

[5.7] Binary Li-Ag brazing alloys have been used to bond a commercially pure titanium honeycomb sandwich structure to a 6Al4V titanium base alloy sheet. Initial tests showed an irregular bonding pattern. It was suspected that the Li (alloyed with Ag as a fluxing agent) was reacting with N_2 trapped in the honeycomb to form tri-lithium nitride, Li_3N, thereby preventing the available lithium from reducing surface oxides and fluxing away surface impurities. Gas analysis revealed that the inert gas used to purge air from the honeycomb contained 0.005 v/o (volume percent) N_2. If brazing is conducted at 800 K, use thermodynamics to prove or disprove the validity of Li_3N formation. State assumptions. Use the free energy data given in Exercise Problem [5.5]. Also see Example Problem 5-5.

> *Ans:* $P_{N2} = 1.15 \times 10^{-10}$ atm; since 0.005 v/o is equivalent to 5×10^{-5} atm, Li_3N will form under experimental conditions. All condensed phases are assumed to be pure.

[5.8] Estimate the temperature at which SiO_2 and Si metal will react to boil off SiO in a vacuum furnace pumped to 10^{-3} atm (personal communication, 1959, R. Schuhmann, Jr., Department of Metallurgical Engineering, Purdue University, West Lafayette, Indiana).

> *Ans:* $T \approx 1511$ K.

[5.9] In Exercise Problem [5.8], it is shown that the best vacuum obtainable is 10^{-3} atm at 1511 K when impure SiO_2 (containing excess Si) is used as containment for heat treatment, annealing, etc. Determine the best vacuum obtainable when stoichiometrically pure SiO_2 is used for containment at the same temperature.

> *Ans:* $P_T = 1.71 \times 10^{-9}$ atm.

[5.10] Refer to Exercise Problem [5.7]. In order to limit the formation of trilithium nitride, the P_{N_2} must be held below 1.15×10^{-10} atm. Since 0.005 v/o (volume percent) N_2 in the inert gas is unacceptable, select a possible candidate metal over which the inert gas could be passed in order to reduce N_2 to an acceptable partial pressure before it is used to purge air prior to brazing.

> *Ans:* Ti(β) and Ca are two possibilities. Ti(β) is preferable.

[5.11] According to Kane and Chakachery (1992), possible coating materials for carbon/carbon composites include oxidation resistant SiC and Si_3N_4 coatings as well as glass forming inhibitors such as boron carbide. The stability of these coatings in hydrogen environments is unknown although data for B_4C and H_2 can be computed from information in the literature. Calculate equilibria for the B, C, H system at 0.03 MPa and 13.8 MPa and at a temperature of 815°C. Discuss.

> *Ans:* At $P_T = 0.03$ MPa = 0.3 atm, $P_{H_2} \approx 0.2999$ atm, $P_{CH_4} \approx 0.0001$ atm; $P_{CH_4}/P_{H_2} \approx 0.0003$. At $P_T = 13.8$ MPa = 136 atm, $P_{H_2} \approx 134.4$ atm, $P_{CH_4} \approx 1.6$ atm; $P_{CH_4}/P_{H_2} \approx 0.012$.

[5.12] Calculate the *maximum* allowable P_{CO_2} needed to stabilize wollastonite at 570°C. Use the data for the contact metamorphic reaction given in Exercise Problem [3.12] and Appendix A, Table A.1. Solid reaction components are in the standard state. State rationale for assumptions used in the calculation.

> *Ans:* $P_{CO_2} = 1440$ atm. Assumptions: $\Delta C_p = 0$ and no correction for nonideal gas behavior.

[5.13] Illustrated below is a *weathering* rind: a rim of chemically weathered and discolored rock enclosing an unweathered rock core.

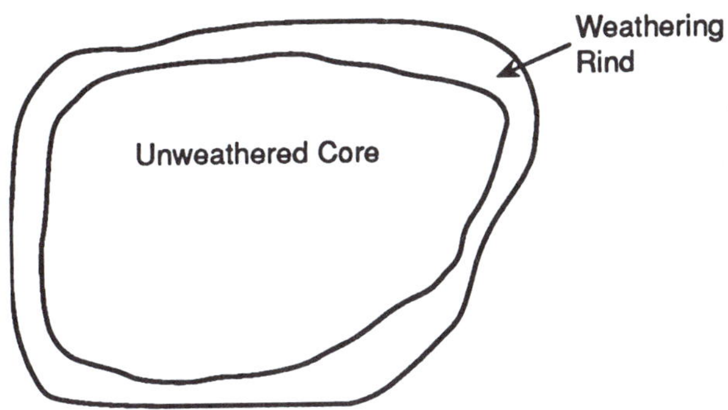

FeO(s), one of the initial chemical weathering products of Fe-rich minerals, frequently undergoes a combination of oxidation and hydration to produce the orthorhombic mineral goethite, FeO(OH). Goethite, a commercial source of iron, imparts a yellow-brown color to weathering rinds. FeO(OH) is also reported as a corrosion product on iron (Pourbaix diagram, *Metals Handbook*, v. 13, p. 38).

(a) Calculate $P_{O_2(g)}$ required to form and stabilize goethite at 298 K. Condensed reaction components are in the standard state.

 Ans: $P_{O_2(g)} = 10^{-84}$ atm $\Rightarrow$ goethite is stable in air at the surface of the earth where $P_{O_2} \approx 0.22$ atm.

(b) What assumption, if any, about the stability of goethite is inherent in the result from (a)?

 Ans: See problem statement.

[5.14] In the absence of sufficient H_2O, chemical weathering of a rock containing iron-rich minerals will produce hematite, Fe_2O_3, instead of goethite (refer to Exercise Problem [5.13]). Hematite imparts a red to reddish-brown color to rocks.

(a) Calculate the minimum $P_{O_2(g)}$ required to form and stabilize hematite at 298 K.

 Ans: $P_{O_2(g)} = 5 \times 10^{-92}$ atm.

(b) Discuss the significance of (a) with respect to the $P_{O_2(g)}$ in air and the stability of hematite at the surface of the earth.

 Ans: See problem statement.

[5.15] Chemical analysis of the reddish and yellow-brown weathering rind on a rock reveals that it contains both goethite and hematite. Based on the results of Exercise Problems [5.13] and [5.14], suggest how the above mineral assemblage might occur.

Ans: (1) Localized area(*s*) of rock depleted in H_2O, hence the mixture of minerals, (2) Zones of H_2O depletion vary over time and promote dissociation of goethite into hematite and H_2O; (3) Changes in pH may also be a factor according to the Pourbaix Diagram (see Exercise Problem [5.13]).

[5.16] Brucite, $Mg(OH)_2$, is slowly heated in an atmosphere where the isobaric partial pressure of H_2O vapor is $10^{1/2}$ atm. At what temperature will Brucite begin to decompose?

> *Ans: T = 584 K.*

[5.17] Methane is produced when water saturated sedimentary rocks containing graphite are heated to elevated temperatures. Calculate the equilibrium $P_{CO_2} \cdot P_{CH_4} / P_{H_2O}^2$ ratio for the reaction

$$2C(s) + 2H_2O(g) \rightarrow CH_4(g) + CO_2(g)$$

at $T = 600°C$ and $P_T = 2$ atm. What is the effect of an increase in P_T? Explain.

> *Ans:* $P_{CO_2} \cdot P_{CH_4} / P_{H_2O}^2 = 0.21$. P_T has no effect on the reaction.

[5.18] Using the ratio computed in Exercise Problem [5.17], calculate the partial pressures of each gas at $600°C$ and $P_T = 2$ atm.

> *Ans:* $P_{H_2O} = 1.044$ atm, $P_{CH_4} = 0.478$ atm; $P_{CO_2} = 0.478$ atm.

[5.19] The mineral siderite, $FeCO_3$, is found in low-temperature hydrothermal veins and low oxygenated continental waters where it is associated with clay and carbonaceous materials. Calculate the P_{CO_2} and P_{CO} required to equilibrate siderite and magnetite, Fe_3O_4, at 300 K in a CO_2-CO atmosphere containing 66 2/3 mole percent CO_2.

> *Ans:* $P_{CO_2} = 12.48 \times 10^{-13}$ atm; $P_{CO} = 6.24 \times 10^{-13}$ atm. Actual partial pressures may be modified by: (1) hydrothermal volatiles associated with igneous processes or (2) decomposing organic matter.

[5.20] Thorium metal is used as an alloying addition in magnesium technology and is also used as a deoxidant for molybdenum, iron, and other metals. However, thorium may corrode in water vapor to form thorium oxide and hydrogen and lose weight by spallation. Discuss the feasibility of controlling the gas phase composition as a method for minimizing corrosion. The following data is provided (Wilkinson and Murphy, 1958, p. 213-215):

$$Th + 2H_2O(g) \rightarrow ThO_2 + 2H_2(g)$$
$$\Delta G_{298}^0 = -171 \text{ kcal.}$$

> *Ans:* $P_{H_2O} / P_{H_2} \approx 2 \times 10^{-63}$. It would be impossible to hold P_{H_2O} low enough to shift the equilibrium from right to left.

CHAPTER 6

HETEROGENEOUS EQUILIBRIA: VARIABLE COMPOSITION IN CONDENSED PHASES

6.1 GAS PARTIAL PRESSURE AND FUGACITY

TL analysis of equilibrium systems involving gases has thus far been restricted to consideration of ideal gas behavior. Although such behavior does not exist at higher pressures, many gases do approach ideal behavior at pressures less than approximately 5 atmospheres. Hence, the previous equations developed are reliable for the analysis of gaseous systems at low pressures. If a pure gas or gas mixture deviates from ideal behavior, it is referred to as a *nonideal* or *real gas*. Because the behavior of a real gas cannot be described by [1-1], the equation of state is modified. Real gas behavior implies that the relationship between ΔG_i and $\ln(P_i)$, expressed by [5-7], is no longer valid. In order to retain linearity, a new thermodynamic property, *fugacity*, is substituted for pressure into the differential form of [5-7]. The fugacity, f_i, of gas component i is defined by

$$dG_i = VdP = RTd\ln (f_i) \qquad [6\text{-}1]$$

where $f_i \rightarrow P_i$ as $P_i \rightarrow 0$. With few exceptions, applications discussed in this book involve low pressures, hence $f = P$ and [5-7] is applicable. For typical discussions of gases at elevated pressures, the reader is referred to Hamill et al. (1966, p. 135–139) and Gaskell (1981, p. 188–205).

6.2 THERMODYNAMIC ACTIVITY

Assuming ideal vapor, the *activity* of a component i is defined by

$$a_i = \frac{f_i}{f_i^0} = \frac{P_i}{P_i^0} \qquad [6\text{-}2]$$

where P_i is the vapor pressure of i over i in condensed solution and P_i^0 and f_i^0 are the vapor pressure and fugacity, respectively, of i over pure condensed i at the same temperature. Note that P_i^0 is normally defined for i in the same state of aggregation, solid or liquid, as the solution. Substituting [6-2] into [5-5],

$$G_i - G_i^0 = RT \ln(a_i) \qquad [6\text{-}3]$$

where G_i and G_i^0 are, respectively, the Gibbs free energy of i in condensed solution and the standard state Gibbs free energy of pure condensed i. Note that [6-3] reduces to [5-6] or [5-7] for a mixture of gases only since $P_i^0 = 1.0$ and $a_i = Y_i P_T = P_i$. Incorporating [6-2] into a typical TL involving condensed state components A and B:

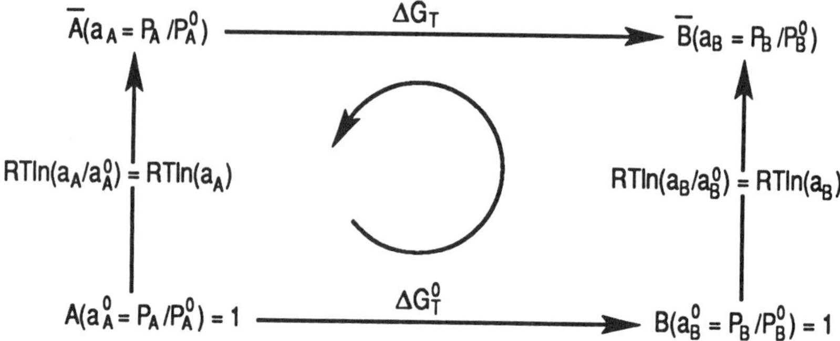

where $\overline{A}$ and $\overline{B}$ denote components A and B, respectively, in solution. In addition, a_A^0 and a_B^0 denote the activity of pure A and B respectively. Summing counter-clockwise,

$$\Sigma\Delta G_{TL} = 0 = \Delta G_T^0 + RT\ln(a_B) - \Delta G_T - RT\ln(a_A)$$

or

$$\Delta G_T = \Delta G_T^0 + RT\ln(a_B/a_A) \qquad [6\text{-}4]$$

The equilibrium constant follows from [6-4], as will be discussed in Chapter 8.

A review of activity and standard states of gas phases is presented in terms of Systems I and II below. The results are summarized in Figures 6.1 and 6.2 and in Table 6.1.

System I: Gas Insoluble in Condensed Phase

Figure 6.1 characterizes an ideal gas, i, which is insoluble in any condensed phase with which it is in contact. Gas i may be either pure or a component in an ideal gas mixture. Since $P_i^0 = 1$ atm at *any temperature* (Section 5.1), the activity of i from [6-2] is $a_i = P_i/1 = P_i$. Using [5-4], $a_i = P_i/1 = Y_i P_T$. Note that if the total gas pressure in Figure 6.1 is $P_T = 1$ atm, then $a_i = P_i = Y_i$.

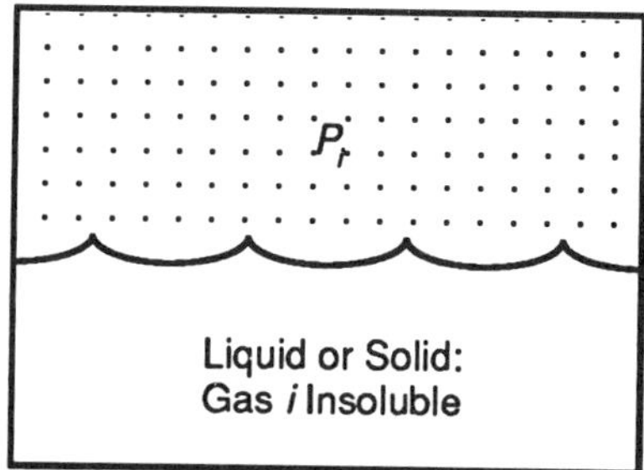

Figure 6.1 Activity of an ideal gas, i, insoluble in the condensed phase. Gas i is either pure or a component in an ideal gas mixture. $P_i^0 \doteq 1$ atm at any temperature, thus $a_i = P_i$

Example Problem 6-1

Sixty moles of an ideal gas mixture at 5 atmospheres pressure contains 15 moles of $S_2(g)$ in contact with microcrystalline quartz, $SiO_2(s)$. If analysis of the quartz reveals that it contains no sulphur impurities, calculate the activity of $S_2(g)$.

Solution

Referring to Table 6.1, $P^0_{S_2(g)} = 1$ atm, since $S_2(g)$ is insoluble in $SiO_2(s)$. Substituting [5-4] into [6-2],

$$a_{S_2(g)} = P_{S_2(g)}/1 \text{ atm} = Y_{S_2(g)}P_T = (15/60)(5).$$

Hence,

$$a_{S_2(g)} = \underline{1.25}.$$

The reader should note that activity is dimensionless and that $a_{S_2(g)} > 1$.

System II: Vapor Soluble in Condensed Phase

Figure 6.2(a) characterizes a vapor, i, in equilibrium with pure condensed i at temperature T. Since $P_i = P_i^0$, the activity of i in the vapor or condensed state is given by $a_i = P_i^0 / P_i^0 = 1$. It then follows from [5-5] that $\Delta G_i = RT \ln(1) = 0$.

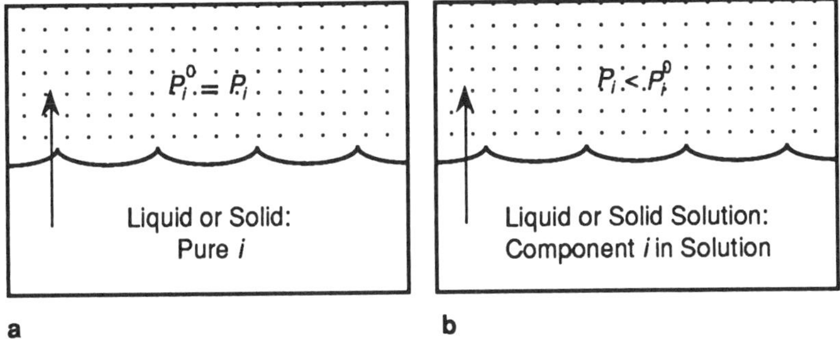

Figure 6.2 Activity of ideal gas component i. (a) Vapor i in equilibrium with pure condensed i at temperature T. (b) Vapor i is soluble in and in equilibrium with a liquid or solid solution which contains i in solution at temperature T.

Figure 6.2(b) characterizes a vapor, i, in equilibrium with component i dissolved in a condensed phase at temperature T. At temperature T, vapor i will exert a pressure less than it would exert if the condensed phase were pure i, hence $P_i < P_i^0$. The *activity of* i *in the vapor*, $a_i^v = (P_i / P_i^0)^v$, is equal to the activity of i in the condensed state, hence $a_i^{cond.} = (P_i / P_i^0)^v$. This equality will be formally derived in Section 7.3.

Table 6.1 Activity in Ideal Gas/Vapor-Condensed Phase Systems

Thermodynamic System	Standard Pressure*	Activity Constant Temperature
System I Gas i (pure or in a gas mixture) is insoluble in condensed phase.**	$P_i^0 = 1$ atm (Any Temp.)	$a_i = P_i$
System II Vapor i in equilibrium with pure condensed i. or	$P_i^0 = $ Vapor Pressure of Pure i.	$a_i = 1$
Vapor i in a gas mixture, i soluble in liquid or solid solution.	$P_i^0 = $ Vapor Pressure of Pure i.	$a_i = P_i / P_i^0$

* The term vapor pressure refers to the partial pressure exerted by a gas component i over a condensed phase containing that component. The term partial pressure refers to a specific component in a gas mixture.

** For gas i soluble in condensed phases, see Example Problem 6–9.

Example Problem 6-2

Calculate the activity of zinc in a liquid copper-zinc alloy if the vapor pressure of $Zn(g)$ over the alloy is 1.28 atm at 1060°C. The vapor pressure of $Zn(g)$ over pure liquid zinc is 4 atm at 1060°C (Darken and Gurry, 1953, p. 412).

Solution

Referring to Table 6.1, $\qquad a_{Zn}^l = P_{Zn(g)} / P_{Zn(g)}^0$.

Hence, $\qquad a_{Zn}^l = 1.28$ atm/4 atm = <u>0.32</u>.

Example Problem 6-3

It may be necessary occasionally to convert concentration units to atomic or mole fraction since these are the basic concentration units used in thermodynamic expressions. Develop expressions for the conversion between w/o and a/o using dimensional consistency. Use m and M to designate mass and molar mass respectively.

Solution
For a binary A-B solution,
$$(\text{a/o})_A = X_A \times 100 = [n_A/(n_A + n_B)] \times 100$$
$$= [m_A/M_A/(m_A/M_A + m_B/M_B)] \times 100.$$
or

$$(\text{a/o})_A = \left(\text{w/o}\big|_A / M_A\right)\Big/\left(\text{w/o}\big|_A / M_A + \text{w/o}\big|_B / M_B\right) \times 100 \qquad [6\text{-}5a]$$

Also,
$$(\text{w/o})_A = [m_A/(m_A + m_B)] \times 100$$
$$= [n_A M_A/(n_A M_A + n_B M_B)] \times 100.$$
Multiplying numerator and denominator by $100/(n_A + n_B)$ and incorporating $\text{a/o}\big|_A$,

$$(\text{w/o})_A = \left[\left(\text{a/o}\big|_A \times M_A\right)\Big/\left(\text{a/o}\big|_A \times M_A + \text{a/o}\big|_B \times M_B\right)\right] \times 100 \qquad [6\text{-}5b]$$

6.3 SOLUTIONS AND PARTIAL MOLAR PROPERTIES

A *solution* is a combination of *two or more components* physically combined so as to form a *homogeneous* mixture. Solution components are elements or compounds that constitute the ingredients of the solution. A *homogeneous solution* is a solid, liquid, or gas that exhibits the same physical and chemical properties throughout. Barring chemical reactions, gas mixtures are always homogeneous. In a binary system, the less abundant component in a solution is referred to as the *solute*, while the more abundant component is referred to as the *solvent*.

The concept of *partial molar property** is best described graphically, using volume. The definitions, based on the graphical representations below, apply to all thermodynamic properties in homogeneous binary A-B systems and may be extended to multicomponent systems. However, such systems are not, in general, discussed in this book. As a result, the designation i for component i will be replaced by an A or B in the binary A-B system. Partial molar properties are listed in SYMBOLS FOR MOLAR PROPERTIES: CHARACTERIZED BY VOLUME.

The relationship between the partial molar properties of binary solution components A and B is illustrated in Figure 6.3. As shown, *ideal* and *nonideal solutions* are characterized by linear and non-linear line segments respectively. Consider a homogeneous material of composition W for which there is complete solubility of A and B in the condensed state. Drawing a tangent line through W, a point on the volume curve for a nonideal solution, and using similar triangles as described by Darken and Gurry (1953, p. 240–241):

---

* The term *partial molal* property is also used, usually by chemists.

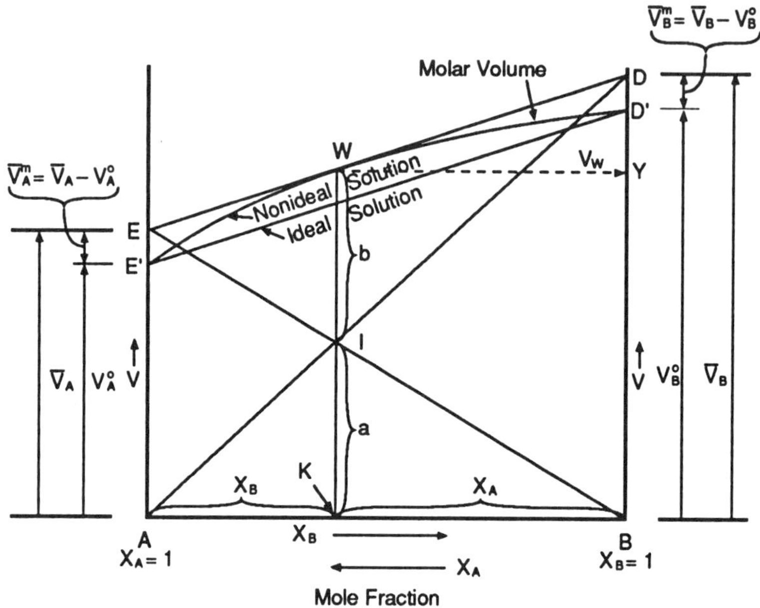

Figure 6.3 Graphical representation of partial molar properties characterized by volume. A and B are completely soluble in all proportions in the condensed state.

$$a/\overline{V}_A = X_A/1.0 \Rightarrow a = X_A \overline{V}_A \qquad \text{(Triangles } BKI \text{ and } BAE\text{)}$$
$$b/\overline{V}_B = X_B/1.0 \Rightarrow b = X_B \overline{V}_B \qquad \text{(Triangles } EWI \text{ and } EDB\text{)}$$

The *molar volume of the solution* at *W* is

$$V_W = a + b = X_A \overline{V}_A + X_B \overline{V}_B \qquad [6\text{-}6]$$

$\overline{V}_A$ and $\overline{V}_B$, located at the intersection of the tangent line at *W* with the vertical axes at the ends of the figure, are the *partial molar volumes* of solution components A and B respectively. The partial molar volume of a solution component is the molar volume occupied by that component *in* solution at a particular composition, *W*.

It can also be shown from Figure 6.3 that

$$\overline{V}_B = \overline{BY} + \overline{YD} = V_W + X_A \frac{\overline{YD}}{X_A} = V_W + (1 - X_B)\frac{dV}{dX_B},$$

where $dV/dX_B = \overline{YD}/X_A$ is the slope of the curve at point *W*. The procedure described above for determining $\overline{V}_A$ and $\overline{V}_B$ is called the *method of tangential intercepts*. This method can be used to obtain partial molar values of other solution properties for any composition.

Consider solution component B. If V^0_B designates the *molar volume of pure component* B, the change in the *partial molar volume of B due to mixing,* $\overline{V}_B^m$, is

$$\overline{V}_B^m = \overline{V}_B - V_B^0 \qquad [6\text{-}7]$$

where B is in the *same* state of aggregation (liquid or solid). The magnitude of $\overline{V}_B^m$ is a measure of the *state of non-ideality,* as will be discussed later. Note that if $\overline{V}_B = V_B^0$, then $\overline{V}_B^m = 0$ and the addition of A atoms has *no effect on the molar volume* occupied by B atoms in the solution. $\overline{V}_B$ may alternately be defined by

$$\overline{V}_B = \left(\partial V' / \partial n_B\right)_{P,T,n_A} \qquad [6\text{-}8]$$

where $\overline{V}_B$ reflects the change in the total volume of a binary solution of arbitrary mass when an infinitely small amount of component B is added so that the overall composition of the solution is not altered at constant temperature and pressure. By analogy with [6-8], the following expressions apply for binary systems:

$$\overline{G}_A = (\partial G' / \partial n_A)_{P,T,n_B} \qquad [6\text{-}9]$$

$$\overline{S}_A = (\partial S' / \partial n_A)_{P,T,n_B} \qquad [6\text{-}10]$$

$$\overline{H}_A = (\partial H' / \partial n_A)_{P,T,n_B} \qquad [6\text{-}11]$$

Darken and Gurry (1953, p. 242-243) discussed the validity of equations analogous to [4-8] for partial molar properties at constant composition and temperature. Using

$$\Delta G_A = \Delta H_A - T\Delta S_A \qquad [6\text{-}12]$$

and substituting $\overline{G}_A^m = \Delta G_A$, $\overline{H}_A^m = \Delta H_A$ and $\overline{S}_A^m = \Delta S_A$ into [6-12],

$$\overline{G}_A^m = \overline{H}_A^m - T\overline{S}_A^m \qquad [6\text{-}13]$$

It should be noted that nearly all of the relationships used to define pure substances can be used to define solution properties, since they are also thermodynamic properties. Substituting $\overline{G}_A^m = \overline{G}_A - G_A^0$ from [6-7] into [6-3],

$$\overline{G}_A^m = RT\ln(a_A) \qquad [6\text{-}14]$$

Using the differential $d(1/T) = -dT/T^2$, an equivalent form of [4-30] is used to define the enthalpy term in [6-13]:

$$\overline{H}_A^m = \left[\frac{\partial(\overline{G}_A^m / T)}{\partial(1/T)}\right]_{P,X_A} \qquad [6\text{-}15]$$

Substituting [6-14] into [6-15],

$$\overline{H}_A^m = R\left[\frac{\partial \ln(a_A)}{\partial(1/T)}\right]_{P,X_A} \qquad [6\text{-}16]$$

Using an equivalent form of [4-27], the entropy term in [6-13] is defined by

$$\overline{S}_A^m = -\left(\frac{\partial \overline{G}_A^m}{\partial T}\right)_{P,X_A} = -R\left[\frac{\partial T\ln(a_A)}{\partial T}\right]_{P,X_A}$$

which expanded gives

$$\overline{S}_A^m = -R\ln(a_A) - RT\left[\frac{\partial \ln(a_A)}{\partial T}\right]_{P,X_A} \qquad [6\text{-}17]$$

Substituting [6-7] and its equivalent for A into [6-6], the *total or integral molar volume of mixing* is

$$V^m = X_A \overline{V}_A^m + X_B \overline{V}_B^m \qquad [6\text{-}18]$$

By analogy, the following total molar mixing expressions also apply to binary systems:

$$G^m = X_A \overline{G}_A^m + X_B \overline{G}_B^m \qquad [6\text{-}19]$$

$$H^m = X_A \overline{H}_A^m + X_B \overline{H}_B^m \qquad [6\text{-}20]$$

$$S^m = X_A \overline{S}_A^m + X_B \overline{S}_B^m \qquad [6\text{-}21]$$

Substituting expressions for A and B from [6-14] into [6-19],

$$G^m = RT[X_A \ln(a_A) + X_B \ln(a_B)] \qquad [6\text{-}22]$$

Using an analogous form of [4-8] as in [6-12],

$$G^m = H^m - TS^m \qquad [6\text{-}23]$$

There are two aspects of solution behavior that should be considered. First, it is common to identify behavior in terms of a specific component. As will be seen, for example, solutions approach ideal behavior at high solvent concentrations as $X_{\text{Solvent}} \rightarrow 1.0$. Second, enthalpy and entropy contributions expressed by [6-13] and defined in general by [6-16] and [6-17] are evaluated separately. The enthalpy contribution is associated with bond energy whereas the entropy contribution is associated with structure, as discussed in Section 3.5. Expressions for $\overline{H}_A^m$ and $\overline{S}_A^m$ are determined from solution models appropriate for the materials system in question. For some systems, $\overline{H}_A^m$ and $\overline{S}_A^m$ are available directly from the literature. However, a trial and error procedure may be useful to extract $\overline{H}_A^m$ and $\overline{S}_A^m$ from solubility data or phase diagrams. Three solution models will now be introduced for use in this and subsequent chapters. Note that the superscript "id" is used to designate an *ideal* solution property.

6.4 SOLUTION MODELS

(1) Ideal Solution: Raoult's Law

Ideal solutions are mixtures in which the bond energy between unlike pairs of atoms is the average of bond energies between like pairs. The ideal solution model is rarely observed in condensed systems over an entire compositional range, hence, an important function of this model is to serve as a reference with which to compare nonideal behavior. These comparisons are normally expressed quantitatively as *excess thermodynamic properties* and are discussed in Section 6.5.

Ideal or Raoultian solution behavior is stated in terms of *Raoult's law*: the vapor pressure exerted by a dissolved component A, P_A, in a homogeneous condensed solution is equal to the product of the atomic fraction of A in the solution, X_A, and the vapor pressure of pure A, P_A^0, at the temperature of the solution. Hence,

$$P_A = X_A P_A^0 \qquad [6\text{-}24]$$

Substituting [6-24] into [6-2],

$$a_A = X_A \qquad [6\text{-}25]$$

[6-25] is an *alternate form of Raoult's law* and is illustrated for an ideal A-B solution in Figure 6.4. The line labeled A_{Ideal} is a plot of a_A versus X_A, and the line labeled B_{Ideal} is a plot of a_B versus X_B. Note that the slope of each line is $a/X = 1$, in accordance with [6-25].

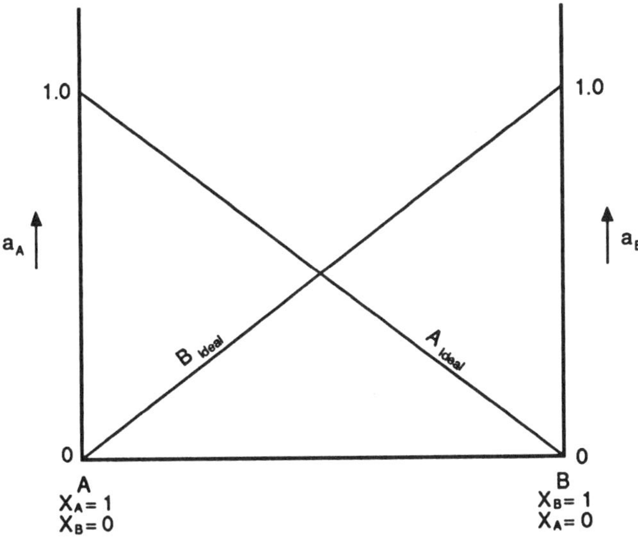

Figure 6.4 Ideal binary A-B solution defined according to Raoult's law, $a_A = X_A$. Solution components A and B are represented by the lines labeled A_{Ideal} and B_{Ideal} respectively. The slopes of these lines are $a_A/X_A = a_B/X_B = 1.0$.

Substituting [6-25] into [6-14],

$$\overline{G}_A^m = RT \ln(X_A) \qquad [6\text{-}26]$$

Substituting [6-25] into [6-16],

$$\overline{H}_A^m = R\left[\frac{\partial \ln(X_A)}{\partial(1/T)}\right]_{P,X_A} \!\!\!\!\!\!\!\!\!\!\!\!\!\nearrow^{0}$$

or

$$\overline{H}_A^m = 0 \tag{6-27}$$

Substituting [6-25] into [6-17],

$$\overline{S}_A^m = -R\ln(X_A) - RT\left[\frac{\partial \ln(X_A)}{\partial T}\right]_{P,X_A} \!\!\!\!\!\!\!\!\!\!\!\!\!\nearrow^{0}$$

or

$$\overline{S}_A^m = -R\ln(X_A) \tag{6-28}$$

Substituting expressions for both A and B from [6-26], [6-27], and [6-28] into [6-19], [6-20], and [6-21] respectively,

$$G^m = RT[X_A \ln(X_A) + X_B \ln(X_B)] \tag{6-29}$$

$$H^m = 0 \tag{6-30}$$

$$S^m = -R[X_A \ln(X_A) + X_B \ln(X_B)] \tag{6-31}$$

A graphical concept of ideality is illustrated in Figure 6.3. If the binary A-B solution of composition W shown were ideal, the curve E'-W-D' would be linear and would follow the trace of the tangent line through W, E-W-D. Note that in such a case, $\overline{V}_A^m = \overline{V}_B^m = 0$.

Example Problem 6-4

Consider a solid solution of components A and B. Assume both components behave ideally. Plot Gibbs partial molar free energy of mixing for both components as well as the total molar free energy of mixing versus mole fraction. Show maxima, minima, and slopes at critical points.

Solution

$\overline{G}_A^m$ and $\overline{G}_B^m$:

 At $X_A = 1$ and $X_B = 0$ and from [6-26]:

$$\overline{G}_A^m = RT\ln(X_A) = RT\ln(1) = 0,$$

$$\frac{d\overline{G}_A^m}{dX_A} = \frac{d[RT\ln(X_A)]}{dX_A} = \frac{RT}{X_A} = \frac{RT}{1} = \text{finite slope;}$$

$$\overline{G}_B^m = RT\ln(X_B) = RT\ln(0) = -\infty.$$

 At $X_B = 1$ and $X_A = 0$:

$$\overline{G}_B^m = RT\ln(1) = 0,$$

$$\frac{d\overline{G}_B^m}{dX_B} = \frac{d[RT\ln(X_B)]}{dX_B} = \frac{RT}{X_B} = \frac{RT}{1} = \text{finite slope;}$$

$$\overline{G}_A^m = RT \ln(0) = -\infty.$$

The limiting slopes are approximated on Figure 6.5.

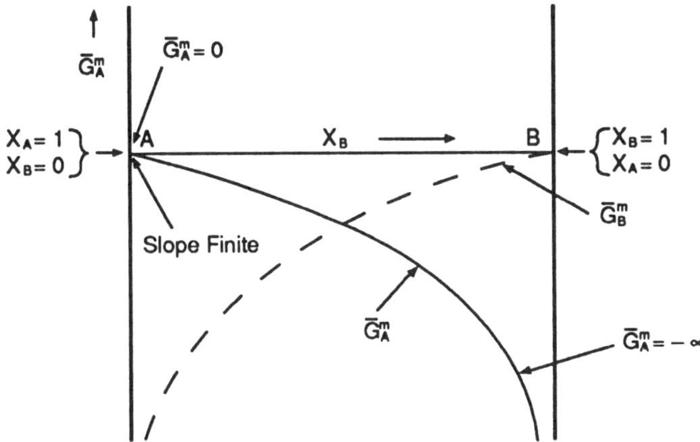

Figure 6.5 $\overline{G}_A^m$ and $\overline{G}_B^m$ versus X_A and X_B, respectively, for a homogeneous ideal binary A-B solution.

$\underline{G^m}$:

   $\underline{At\ X_A = 1\ and\ X_B = 0}$ and from [6-29]:

   $G^m = RT[X_A\ln(X_A) + X_B\ln(X_B)] = RT[(1)\ln(1) + 0\ln(0)] = 0\ln(0)$.
   Since $0\cdot\ln(0) = 0\cdot(-\infty)$ is indeterminate, L'Hopital's rule (Protter and Morrey, 1970, p. 632-635) is used to evaluate G^m:

   Let $Y = \dfrac{G^m}{RT} = X_B \ln(X_B) = \dfrac{\ln(X_B)}{1/X_B} = \dfrac{f(X_B)}{F(X_B)}$

   where $f(X_B) = \ln(X_B)$ and $F(X_B) = 1/X_B$. Then

   $$\lim_{X_B \to 0} \frac{f(X_B)}{F(X_B)} = \lim_{X_B \to 0} \frac{f'(X_B)}{F'(X_B)} = \lim_{X_B \to 0} \frac{(1/X_B)}{(-1/X_B^2)} = \lim_{X_B \to 0}(-X_B) = 0.$$

   Consequently, at $X_B = 0$, $G^m = 0\cdot\ln(0) = 0$. Taking the derivative of [6-29],

   $$\frac{dG^m}{dX_A} = RT[\ln(X_A) - \ln(X_B)] = RT[\ln(1) - \ln(0)] = \infty.$$

   $\underline{At\ X_B = 1\ and\ X_A = 0}$:

   The calculations are analogous to those immediately above:
   $G^m = 0$ and $dG^m/dX_B = -\infty$.

The limiting slopes are approximated on Figure 6.6.

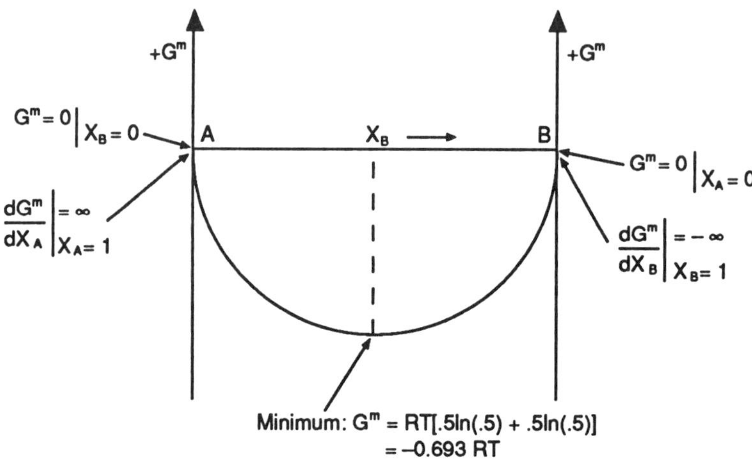

Figure 6.6 G^m versus X_A and X_B for a homogeneous ideal binary A-B solution.

(2) Nonideal Solutions

Nonideal solutions are mixtures in which the attraction between like pairs of atoms is greater than the attraction between unlike pairs (*positive deviation from ideality*) or less than the attraction between unlike pairs (*negative deviation from ideality*). The arrangement of atoms may or may not be random and substitutional. Non-ideality is introduced into [6-25] by inserting the thermodynamic *activity coefficient*, γ, into the expression. Hence, [6-25] becomes

$$a_A = \gamma_A X_A \qquad [6\text{-}32]$$

For ideal solutions, $\gamma_A = 1.0$. In general, γ_A is a function of both temperature and composition, hence

$$[\partial \ln(a_A) / \partial T]_{P,X_A} \neq 0.$$

For γ_A = (constant),

$$[\partial \ln(a_A) / \partial T]_{P,X_A} = 0.$$

(a) *Henrian Solution Model:*

The *Henrian solution model* applies to mixtures in which a solute, A, is at low concentration, usually $X_A < 1$ a/o. The distance between solute atoms at such concentrations is large, hence thermodynamic properties are additive and in direct proportion to solute concentration. At low solute concentrations, solute behavior in a dilute solution is expressed by *Henry's Law*:

$$X_A = k_A P_A \qquad [6\text{-}33]$$

where X_A and P_A have the usual meaning and k_A is a temperature dependent constant. Substituting [6-32] and [6-33] into [6-2],

$$\gamma_A = \frac{1}{(P_A^0 k_A)} = b \text{ (constant)}.$$

Substituting into [6-32],

$$a_A = bX_A \qquad [6\text{-}34]$$

(b) *Dilute Solution Model-Modified Henrian Solution:*

The *dilute solution model* is essentially identical to the Henrian model and applies to low concentration solutes. The difference relates to how the non-ideality, γ_A, is expressed. For example, substituting [6-34] into [6-14],

$$\overline{G}_A^m = RT \ln(bX_A).$$

Substituting this expression into [6-15] and [6-17],

$$\overline{H}_A^m = \left[\frac{\partial R \ln(bX_A)}{\partial(1/T)} \right]_{P,X_A} = 0;$$

$$\overline{S}_A^m = -R \ln(bX_A) - RT \left[\frac{\partial \ln(bX_A)}{\partial T} \right]_{P,X_A}$$

$$= -R \ln(b) - R \ln(X_A)$$
$$= (\text{constant}) - R \ln(X_A).$$

From the above analysis, it is observed that non-ideality, according to Henry's law, is identified only with the entropy term since enthalpy is zero. Kubaschewski and Alcock (1979, p. 47) comment on the increased accuracy that is achieved when non-ideality is distributed between both enthalpy and entropy terms. Parameters incorporating both terms define the dilute solution model.

The enthalpy and entropy contributions of solute A in the dilute solution model (personal communication, 1959, R. Schuhmann, Jr., Department of Metallurgical Engineering, Purdue University, West Lafayette, Indiana) are described by

$$\overline{H}_A^m = h \text{ (constant)} \qquad [6\text{-}35]$$

$$\overline{S}_A^m = s \text{ (constant)} - R \ln(X_A) \qquad [6\text{-}36]$$

Substituting [6-35] and [6-36] into [6-13],

$$\overline{G}_A^m = h - T[s - R \ln(X_A)]$$
$$= RT \ln(X_A) + h - Ts \qquad [6\text{-}37]$$

Inserting [6-32] into [6-14] and comparing with [6-37] term by term, $RT \ln(\gamma_A) = h - Ts$. Solving for γ_A,

$$\gamma_A = b = \exp\left(\frac{h - Ts}{RT} \right) \qquad [6\text{-}38]$$

where b is constant at constant temperature. The Henrian and modified Henrian (dilute) solution models are summarized in Table 6.2.

Table 6.2 Henrian and Modified Henrian (Dilute) Solution Models*

Solution Model	$\overline{H}_A^m$	$\overline{S}_A^m$
Henrian Solution	0	$s - R\ln(X_A)$
Dilute Solution (Modified Henrian)	h	$s - R\ln(X_A)$

* h and s are temperature dependent constants but over small temperature intervals may be assumed constant, as in Chapter 7.

Where specific interest is focused at a low concentration to one component, it is convenient to introduce an alternative to the pure component standard state. This alternative defines b such that $b \rightarrow 1.0$ at infinite dilution. For the purposes of this book, the pure component standard state will be used in all calculations.

(3) Regular Solution Model

Regular solutions are comprised of components at intermediate concentrations. Interactions between like and unlike atoms vary with solution composition, hence, γ_i is *variable* with X_i. An example of a regular solution is a binary A-B substitutional solid solution with random arrangement of A and B atoms on A and B sites.

The *quasi-chemical theory* (Swalin, 1964, p. 109-116) relates chemical effects to energetics rather than mechanical or valence effects and demonstrates that the enthalpy associated with a regular solution is *nonideal*. Vibrational effects are unchanged during the solution process, hence, entropy is *ideal*. The *nonideal* partial molar enthalpy of mixing for a binary A-B solution component A is

$$\overline{H}_A^m = \Omega(1 - X_A)^2 \qquad [6\text{-}39]$$

where

$$\Omega = zN_{Av}[h_{AB} - (h_{AA} + h_{BB})/2] \qquad [6\text{-}40]$$

z is the number of bonds (coordination number) per atom and N_{AV} is Avogadro's number. The enthalpies per A-A, B-B, and A-B bond are designated h_{AA}, h_{BB}, and h_{AB} respectively. Ω is *constant and independent of temperature*. Substituting [6-39] into [6-20],

$$H^m = \Omega X_A(1 - X_A)^2 + \Omega X_B(1 - X_B)^2.$$

Since $(1 - X_A) = X_B$ and $(1 - X_B) = X_A$,

$$H^m = \Omega X_A X_B^2 + \Omega X_B X_A^2$$

or

$$H^m = \Omega X_A X_B \qquad [6\text{-}41]$$

Since a regular solution is ideal with respect to entropy,

$$\overline{S}_A^m = -R \ln(X_A) \qquad [6\text{-}42]$$

Substituting [6-39] and [6-42] into [6-13],

$$\overline{G}_A^m = \Omega(1 - X_A)^2 + RT \ln(X_A) \qquad [6\text{-}43]$$

Consider now a plot of H^m versus composition according to [6-41] and shown in Figure 6.7. If $\Omega > 0$, $H^m > 0$, mixing is endothermic, the solution exhibits a positive deviation from ideality, and like pairs of atoms attract. If $\Omega = 0$, the solution is ideal. If $\Omega < 0$, $H^m < 0$, mixing is exothermic, the solution exhibits a negative deviation from ideality, and unlike pairs of atoms attract. For further reading refer to DeHoff (1993, p. 196-203) and Gaskell (1981, p. 366-373).

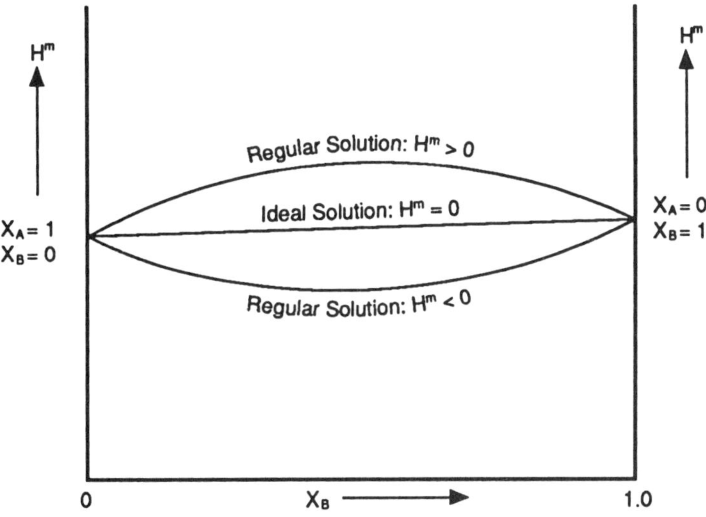

Figure 6.7 Positive ($H^m > 0$) and negative ($H^m < 0$) departures from ideality ($H^m = 0$) for ideal and regular binary A-B solutions. Note that positive and negative deviations apply in general to all nonideal solutions.

6.5 EXCESS THERMODYNAMIC PROPERTIES AND ALTERNATE SOLUTION MODELS

The difference between a nonideal and ideal thermodynamic property is defined as an *excess* of that property. Excess properties, sometimes referred to as *residuals*, are point functions and can be treated as exact differentials. There are occasions in which excess properties are easier to analyze mathematically than the properties themselves. Superscript "XS" is used to designate an excess property. Hence, for component A,

$$\overline{G}_A^{XS} = \overline{G}_A^m - \overline{G}_A^{m,\mathrm{id}} \qquad [6\text{-}44]$$

$$\overline{H}_A^{XS} = \overline{H}_A^m - \overline{H}_A^{m,\mathrm{id}} \qquad [6\text{-}45]$$

$$\overline{S}_A^{XS} = \overline{S}_A^m - \overline{S}_A^{m,\mathrm{id}} \qquad [6\text{-}46]$$

where superscript "id" refers to the ideal solution. Using an analogous form of [4-8] as in [6-12],

$$\overline{G}_A^{XS} = \overline{H}_A^{XS} - T\overline{S}_A^{XS} \tag{6-47}$$

Substituting [6-32] and [6-25] into [6-44],

$$\overline{G}_A^{XS} = RT\ln(\gamma_A) + RT\ln(X_A) - RT\ln(X_A)$$

or

$$\overline{G}_A^{XS} = RT\ln(\gamma_A) \tag{6-48}$$

Additional relationships can now be developed for each of the solution models discussed.

(1) Ideal Solution ($\gamma_A = 1$)

From [6-26] and [6-44] or from [6-48],

$$\overline{G}_A^{XS} = RT\ln(X_A) - RT\ln(X_A) = RT\ln(1) = 0.$$

Substituting [6-27] into [6-45] and [6-28] into [6-46],

$$\overline{H}_A^{XS} = 0;$$

$$\overline{S}_A^{XS} = 0.$$

In general, any excess thermodynamic property for a solution component exhibiting ideal behavior is zero.

(2) Dilute Solution (γ_A is constant $\neq 1$)

Substituting [6-26] and [6-37] into [6-44] and comparing the result with [6-47],

$$\overline{H}_A^{XS} = h;$$
$$\overline{S}_A^{XS} = s.$$

(3) Regular Solution (γ_A is variable)

Substituting [6-26], [6-43], and [6-48] into [6-44],

$$\overline{G}_A^{XS} = \Omega(1 - X_A)^2 + RT\ln(X_A) - RT\ln(X_A) = \Omega(1 - X_A)^2 = RT\ln(\gamma_A);$$

$$\gamma_A = \exp[\Omega(1 - X_A)^2/RT] \tag{6-49}$$

Note that if Ω is positive, $\gamma_A > 1$ and like pairs of atoms attract. If Ω is negative, $\gamma_A < 1$ and unlike pairs attract.

(4) Alternate Solution Models

In this book, ideal, dilute, and regular solution models are applied analytically. Variations in these models incorporate excess entropy. Two alternate models are introduced here but are not used in any problem solutions.

(a) *Athermal Solution:*

These solutions are characterized by large size differences between component atoms. Solutions containing Fe and Na approach this type of behavior, as noted by Johnson (1964). Internal energy in athermal solutions is governed almost entirely by excess entropy. Hence, the following expressions define enthalpy and entropy contributions:

$$\overline{H}_A^{XS} = 0;$$

$$\overline{S}_A^{XS} = s \text{ (constant)}$$

or

$$\overline{S}_A^{XS} = \delta(1 - X_A)^2$$

where δ is constant.

Combining [6-13], [6-45], and [6-46] and substituting for $\overline{S}_A^{XS}$,

$$\overline{G}_A^m = T[R \ln(X_A) - \delta(1 - X_A)^2] \qquad [6\text{-}50]$$

(b) *Subregular Solutions*:

Internal energy is governed by both size and chemical bond effects, hence characteristics of both athermal and regular solutions are incorporated into the model. The following expressions define enthalpy and entropy contributions where Ω and ω are constants. Substituting as above,

$$\overline{H}_A^{XS} = \Omega(1 - X_A)^2;$$

$$\overline{S}_A^{XS} = \omega(1 - X_A)^2.$$

The Gibbs partial molar free energy of mixing of A becomes

$$\overline{G}_A^m = \Omega(1 - X_A)^2 - T[\omega(1 - X_A)^2 - R \ln(X_A)] \qquad [6\text{-}51]$$

6.6 TL ANALYSIS OF SOLUTIONS IN HETEROGENEOUS REACTIONS

Example Problem 6-5

The following enthalpy data are given for Au-Cu alloys at a temperature of 500°C (Swalin, 1964, p. 329). H^m is in cal/mol.

X_{Cu}	0.1	0.2	0.3	0.4	0.5	0.6	0.7	0.8	0.9
H^m	−355	−655	−910	−1120	−1280	−1240	−1130	−860	−460

(a) Examine the data and identify the solution model most appropriate for this system.

Solution

The solution is not ideal in the range of compositions given since $H^m \neq 0$. Because of the wide range of compositions for X_{Cu}, the data is best analyzed using the regular solution model.

(b) Find expressions for H^m and $\overline{H}_{Cu}^m$.

Solution

Using [6-41], values of Ω are computed at each composition. The results are recorded in the table below:

X_{Cu}	0.1	0.2	0.3	0.4	0.5	0.6	0.7	0.8	0.9
X_{Au}	0.9	0.8	0.7	0.6	0.5	0.4	0.3	0.2	0.1
$X_{Cu}X_{Au}$	0.09	0.16	0.21	0.24	0.25	0.24	0.21	0.16	0.09
H^m	−355	−655	−910	−1120	−1280	−1240	−1130	−860	−460
Ω	−3944	−4094	−4333	−4667	−5120	−5167	−5381	−5375	−5111

Rigorous application of [6-41] would require that Ω remain constant. A random variation would be acceptable for engineering applications. Noting that Ω increases to $X_{Cu} = 0.7$ and then decreases, assume an average $\Omega = -4800$ cal/mol $\times$ (4.184 J/cal) = −20,083 J/mol. Hence, from [6-41],

$$H^m = -4800\, X_{Cu}X_{Au} \text{ cal/mol}$$
$$= -20{,}083 X_{Cu} X_{Au} \text{ J/mol.}$$

From [6-39],

$$\overline{H}^m_{Cu} = -4800(1 - X_{Cu})^2 \text{ cal/mol}$$
$$= -20{,}083(1 - X_{Cu})^2 \text{ J/mol.}$$

(c) Determine G^m at $X_{Cu} = 0.45$.

Solution

Using [6-41],

$$H^m = -20{,}083\, X_{Cu}X_{Au} = -20{,}083 \times 0.45 \times 0.55$$
$$= -4971 \text{ J/mol.}$$

Substituting [6-42] for both components into [6-21],

$$S^m = -R[X_A \ln(X_A) + X_B \ln(X_B)] \qquad [6\text{-}52]$$
$$S^m = -8.3144[0.45\ \ln(0.45) + 0.55\ \ln(0.55)] = 5.72 \text{ J/(mol·K).}$$
$$G^m = -4971 - (773)(5.72) \Rightarrow G^m = \underline{-9393 \text{ J/mol.}}$$

(d) Determine the activity coefficient of Cu, γ_{Cu}, at $X_{Cu} = 0.45$.

Solution

From [6-49],

$$\gamma_{Cu} = \exp[-20{,}083(1 - 0.45)^2/8.3144 \times 773] \Rightarrow \gamma_{Cu} = \underline{0.389}.$$

(e) Find the partial pressure of Cu, P_{Cu}, at $X_{Cu} = 0.45$.

Solution

From [6-2] and [6-32],

$$a_{Cu} = P_{Cu}/P^0_{Cu} = \gamma_{Cu}X_{Cu} \Rightarrow P_{Cu} = P^0_{Cu}\, \gamma_{Cu}X_{Cu}.$$

Hence,
$$P_{Cu} = 0.389 \times 0.45\, P_{Cu}^0 = 0.175\, P_{Cu}^0.$$
From Appendix A, Table A.5,
$$Log_{10}[\,P_{Cu(s)}^0\,(mm\ Hg)] = (-17{,}770/773) - 0.86\log_{10}(773) + 12.29$$
$$= -13.182 \Rightarrow P_{Cu}^0 = 6.58 \times 10^{-14}\ mm\ Hg.$$
Therefore,
$$P_{Cu} = 0.175 \times 6.58 \times 10^{-14}$$
$$= 1.15 \times 10^{-14}\ mmHg.$$

(f) Calculate the $P_{O_2(g)}$ in equilibrium with the Cu-Au alloy for $X_{Cu} = 0.45$ at 500°C. Assume $a_{Cu_2O(s)} = 1.0$ [pure cuprite, $Cu_2O(s)$].

Solution

(1) Set Up. A horizontal line above a condensed phase is used to indicate that the component is in solution.

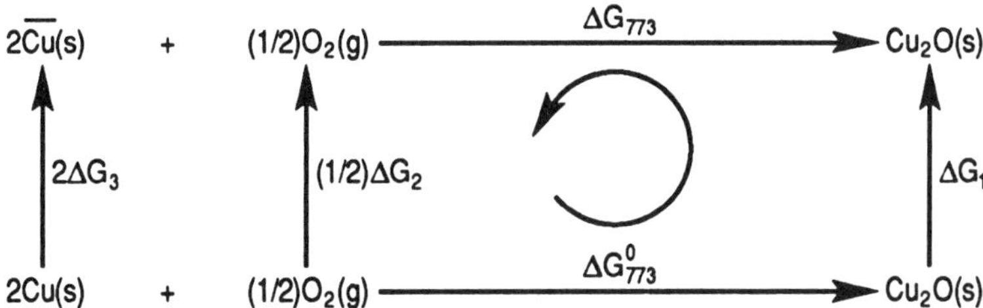

(2) Sum.
$$\Sigma\Delta G_{TL} = 0 = \Delta G_{773}^0 + \Delta G_1 - \Delta G_{773} - (1/2)\Delta G_2 - 2\Delta G_3.$$

(3) Substitute.

$\Delta G_1 = RT \ln(1) = 0$ and $\Delta G_{773} = 0$ at equilibrium.

From Table A.4,
$$\Delta G_T^0 = -169{,}470 - 16.40\,T\log(T) + 123.44(T)\ \text{from 298 to 1356 K} \Rightarrow$$
$$\Delta G_{773}^0 = -169{,}470 - 16.40(773)\log(773) + 123.44(773)$$
$$= -110{,}665\ J/mol.$$
$$(1/2)\Delta G_2 = (1/2)RT \ln(P_{O_2}).$$
$$2\Delta G_3 = 2\overline{G}_{Cu}^m = 2RT \ln(a_{Cu}) = 2(8.3144)(773)\ln(0.389 \times 0.45)$$
$$= -22{,}401\ J/mol.$$
Substituting the above data into $\Sigma\Delta G_{TL} = 0$,
$$0 = -110{,}665 - (1/2)RT \ln(P_{O_2}) - (-22{,}401).$$

(4) Solve.

$$(1/2)RT \ln(P_{O_2}) = -110{,}665 + 22{,}401 = -88{,}264.$$
$$\ln(P_{O_2}) = 2(-88{,}264)/(8.3144 \times 773) \Rightarrow$$
$$P_{O_2(g)} = \underline{1.18 \times 10^{-12}\ atm.}$$

Example Problem 6-6

(a) Will a gas mixture containing 97 v/o H_2O vapor and 3 v/o $H_2(g)$ deoxidize nickel oxide over pure nickel at 1000 K? Assume $P_T = 1$ atm (Darken and Gurry, 1953, p. 513).

Solution

(1) Set Up.

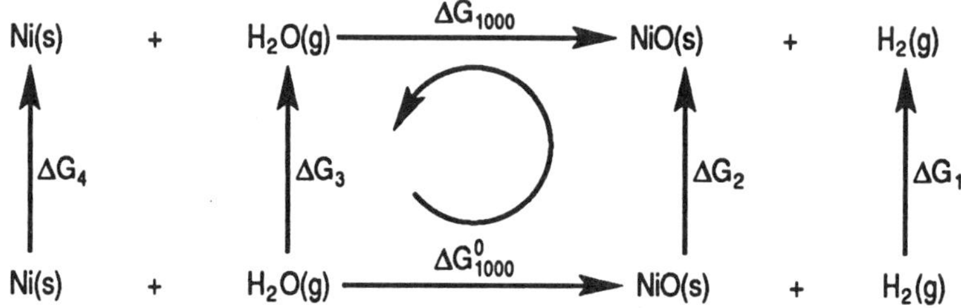

(2) Sum.

$$\Sigma \Delta G_{TL} = 0 = \Delta G^0_{1000} + \Delta G_1 + \Delta G_2 - \Delta G_{1000} - \Delta G_3 - \Delta G_4.$$

(3) Substitute. From [4-9],

$$\Delta G^0_{1000} = \Sigma n \Delta G^0_{1000}(\text{Products}) - \Sigma n \Delta G^0_{1000}(\text{Reactants}).$$

$\Delta G^0_{1000} = \Delta G^0_{1000,NiO(s)} - \Delta G^0_{1000,H_2O(g)}$. From Table A.4,

$$\Delta G^0_{1000} = -244,580 + 98.54(1000) - [-246,460 + 54.82(1000)]$$
$$= 45,600 \text{ J/mol.}$$

$\Delta G_1 = RT \ln(P_{H_2}) = 8.3144(1000)\ln(P_{H_2}) = 8314.4 \ln(P_{H_2}).$

From [5-4], $P_{H_2} = Y_{H_2} P_T = 0.03(1 \text{ atm}) = 0.03$ atm.

Hence, $\Delta G_1 = 8314.4 \ln(0.03)$.

$\Delta G_2 = RT \ln(1) = 0.$

$\Delta G_3 = RT \ln(P_{H_2O}) = 8.3144(1000)\ln(P_{H_2O}) = 8314.4\ln(P_{H_2O}).$

From [5-4], $P_{H_2O} = 0.97(1 \text{ atm}) = 0.97$ atm. Hence,

$\Delta G_3 = 8314.4 \ln(0.97)$.

$\Delta G_4 = \overline{G}^m_{Ni} = 0$ (pure solid).

Substituting the above data into $\Sigma \Delta G_{TL} = 0$,

$$0 = 45,600 + 8314.4\ln(0.03) - \Delta G_{1000} - 8314.4\ln(0.97).$$

(4) Solve.

$\Delta G_{1000} = 45,600 + 8314.4 \ln(0.03/0.97) = 16,698$ J/mol.

Since $\Delta G_{1000} > 0$, NiO(s) will tend to deoxidize.

Figure 6.8, a simplified portion of Figure E.2, can be used to confirm that the reaction will be deoxidizing. From the nomograph, $P_{H_2}/P_{H_2O} \approx 10^{-2} = 0.02$. Since the actual ratio $P_{H_2}/P_{H_2O} = 0.03/0.97 = 0.031$, the reaction will tend to shift from right to left.

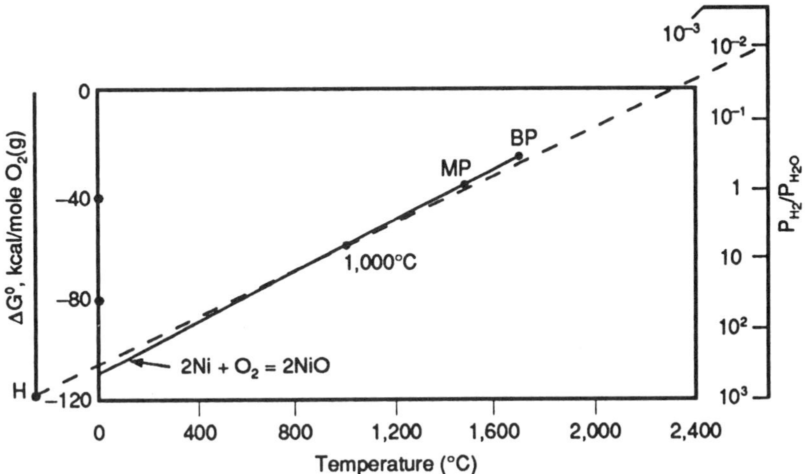

Figure 6.8 Ellingham diagram for Ni oxidation. ΔG^0 versus T (°C) is shown by the solid line. M.P. and B.P. represent the melting and boiling points of Ni(s) respectively. The dashed line, drawn from point H on the vertical line at the left side of the diagram through the 1000°C point to the vertical scale at the right, graphically determines the P_{H_2}/P_{H_2O} ratio. The result is $P_{H_2}/P_{H_2O} \approx 10^{-2}$.

(b) An alloy contains 20 a/o Ni and 80 a/o Au in solid solution at 1000 K. This alloy reacts with water vapor to form NiO(s). Experimental measurement indicates the reaction reaches equilibrium when the H₂O-H₂ mixture con tains 0.35 v/o H₂. Find the activity coefficient of Ni in the alloy. Assume NiO(s) is pure oxide. How does alloying affect the result obtained in part (a)?

Solution

(1) Set Up.

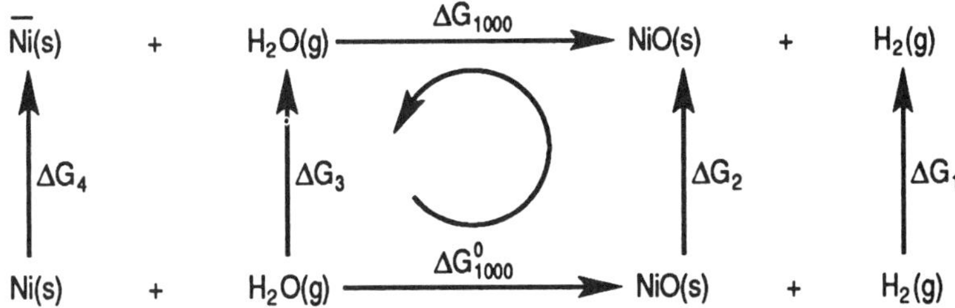

(2) Sum.

$$\Sigma \Delta G_{TL} = 0 = \Delta G^0_{1000} + \Delta G_1 + \Delta G_2 - \Delta G_{1000} - \Delta G_3 - \Delta G_4.$$

(3) Substitute.

$\Delta G^0_{1000} = 45,600$ J/mol, from part (a) above.

$\Delta G_1 = RT \ln(P_{H_2}) = 8.3144(1000) \ln(0.0035)$.

$\Delta G_2 = RT \ln(1) = \Delta G_{1000}$ (equilibrium) $= 0$.

$\Delta G_3 = RT \ln(P_{H_2O}) = 8.3144(1000) \ln(1 - 0.0035)$.

$\Delta G_4 = \overline{G}^m_{Ni} = RT \ln(a_{Ni}) = 8.3144(1000)\ln(a_{Ni})$.

Substituting the above data into $\Sigma\Delta G_{TL} = 0$,

$0 = 45,600 + 8314.4 \ln(0.0035) - 8314.4 \ln(0.9965) - 8314.4 \ln(a_{Ni})$.

(4) Solve.

$a_{Ni} = 0.846 = \gamma_{Ni}X_{Ni} \Rightarrow \gamma_{Ni} = 0.846/0.2 = \underline{4.23}$.

If the effect of alloying is incorporated into part (a),

$\Delta G_4 = \overline{G}^m_{Ni} = 8.3144(1000)\ln(0.846)$.

Therefore,

$\Sigma\Delta G_{TL} = 0 = 45,600 + 8314.4\ln(0.03/0.97) + 1390 - \Delta G_{1000}$

or

$\Delta G_{1000} = \underline{18,088}$ J/mol.

The tendency to deoxidize is increased when Ni is in solid solution with Au since ΔG_{1000} increases.

Example Problem 6-7

Repeat Example Problem 5-2 for the case where Cr forms a 15 a/o solid solution with a metal, M. Assume Cr is more readily oxidized than M and that Cr_2O_3 is insoluble in solution. Assume an ideal Cr-M solid solution. (Darken and Gurry, 1953, p. 513).

Solution
(1) Set Up.

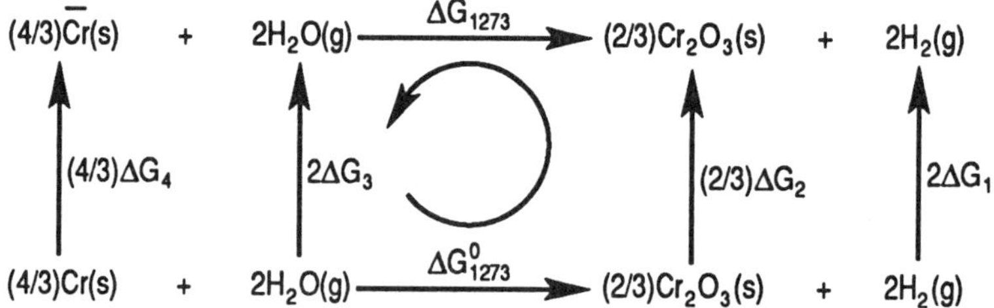

(2) Sum.

$\Sigma\Delta G_{TL} = 0 = \Delta G^0_{1273} + 2\Delta G_1 + (2/3)\Delta G_2 - \Delta G_{1273} - 2\Delta G_3 - (4/3)\Delta G_4$.

(3) Substitute.

$\Delta G^0_{1273} = -173,039$ J, from Example Problem 5-2.

$\Delta G_{1273} = 0$ at equilibrium.

$2\Delta G_1 = 2R(1273)\ln{(P_{H_2})}$ and $2\Delta G_3 = 2R(1273)\ln{(P_{H_2O})}$.

$2/3\Delta G_2 = 2/3RT\ln(1) = 0$.

$4/3\Delta G_4 = 4/3\,\overline{G}^m_{Cr} = 4/3R(1273)\ln(X_{Cr}) = 4/3R(1273)\ln(0.15)$.

Substituting the above data into $\Sigma\Delta G_{TL} = 0$,

$$0 = -173,039 + 2(8.3144)(1273)\ln{(P_{H_2})} - 2(8.3144)(1273)\ln{(P_{H_2O})} -$$
$$4/3(8.3144)(1273)\ln(0.15).$$

(4) Solve.

$$P_{H_2}/P_{H_2O} = \underline{1002}.$$

Comparing this result with that of Example Problem 5-2 for which P_{H_2}/P_{H_2O} $= 3.55 \times 10^3$, alloying reduces the ratio of partial pressures by a factor of nearly 3.5.

6.7 THE GIBBS-DUHEM EQUATION

For a binary A-B solution at constant pressure and temperature,

$$G' = f(n_A, n_B)$$

where n_A and n_B are the number of moles of components A and B respectively. Applying the chain rule and [6-9],

$$dG' = \left(\frac{\partial G'}{\partial n_A}\right)_{P,T,n_B} dn_A + \left(\frac{\partial G'}{\partial n_B}\right)_{P,T,n_A} dn_B$$

or

$$dG' = \overline{G}_A\,dn_A + \overline{G}_B\,dn_B \qquad [6\text{-}53]$$

Replacing V in [6-6] with G and multiplying through by $(n_A + n_B)$,

$$G(n_A + n_B) = X_A(n_A + n_B)\overline{G}_A + X_B(n_A + n_B)\overline{G}_B$$

or

$$G' = \frac{n_A}{(n_A + n_B)}(n_A + n_B)\overline{G}_A + \frac{n_B}{(n_A + n_B)}(n_A + n_B)\overline{G}_B.$$

Hence,

$$G' = n_A\overline{G}_A + n_B\overline{G}_B.$$

Taking the total derivative of this expression,

$$dG' = n_A d\overline{G}_A + \overline{G}_A dn_A + n_B d\overline{G}_B + \overline{G}_B dn_B \qquad [6\text{-}54]$$

Equating [6-53] and [6-54],

$$n_A d\overline{G}_A + n_B d\overline{G}_B = 0 \qquad [6\text{-}55]$$

Dividing [6-55] by $(n_A + n_B)$,

$$X_A d\overline{G}_A + X_B d\overline{G}_B = 0 \qquad [6\text{-}56]$$

Similarly,

$$X_A d\overline{H}_A + X_B d\overline{H}_B = 0 \qquad [6\text{-}57]$$

and

$$X_A d\overline{S}_A + X_B d\overline{S}_B = 0 \qquad [6\text{-}58]$$

In general, for a *solution* containing *any* number of components at *constant pressure and temperature*,

$$\Sigma X_i d\overline{\theta}_i = 0 \qquad [6\text{-}59]$$

or

$$\Sigma n_i d\overline{\theta}_i = 0 \qquad [6\text{-}60]$$

where $\overline{\theta}_i$ is *any partial molar* thermodynamic property of component i. Both [6-59] and [6-60] are equivalent forms of the *Gibbs-Duhem equation*.

Example Problem 6-8

If the solute in a binary A-B condensed solution obeys Henry's law, show that the solvent obeys Raoult's law. Let component A be the solvent and B be the solute. Sketch the result for $A_{solvent}$, B_{solute} and A_{solute}, $B_{solvent}$.

Solution

Substituting the differential form of [6-14], $d\overline{G}_A^m = RT\, d\ln(a_A)$ and $d\overline{G}_B^m = RT\, d\ln(a_B)$, into [6-56] and dividing through by RT,

$$X_A\, d\ln(a_A) + X_B\, d\ln(a_B) = 0 \qquad [6\text{-}61]$$

Substituting [6-34] (component B) into [6-61] and solving for a_A,

$$d\ln(a_A) = -\frac{X_B}{X_A} d\ln(bX_B) = -\frac{X_B}{X_A}\left[d\ln(b) + d\ln(X_B)\right]$$

$$= -\frac{X_B}{X_A} d\ln(X_B).$$

Substituting $X_A = (1 - X_B)$ and integrating over limits that yield $\ln(a_A)$ directly,

$$\int_1^{a_A} d\ln(a_A) = -\int_0^{X_B} \frac{X_B d\ln(X_B)}{1 - X_B}$$

or

$$\ln(a_A) - \ln(1) = \ln(1 - X_B) - \ln(1) \Rightarrow \ln(a_A) = \ln(X_A).$$

Therefore,

$$\underline{a_A = X_A \Rightarrow A \text{ is ideal.}}$$

Figure 6.9 is a plot of activity versus composition for both components. As illustrated in the figure, both Raoultian and Henrian behavior are approached at high concentration and dilution respectively. In this sense, Raoult's and

Henry's law are limiting laws (Lupis, 1983, p. 158–163). The activity (slope) of each line in mid-range compositions is variable and compositionally dependent. The regular solution model is often applicable in this range.

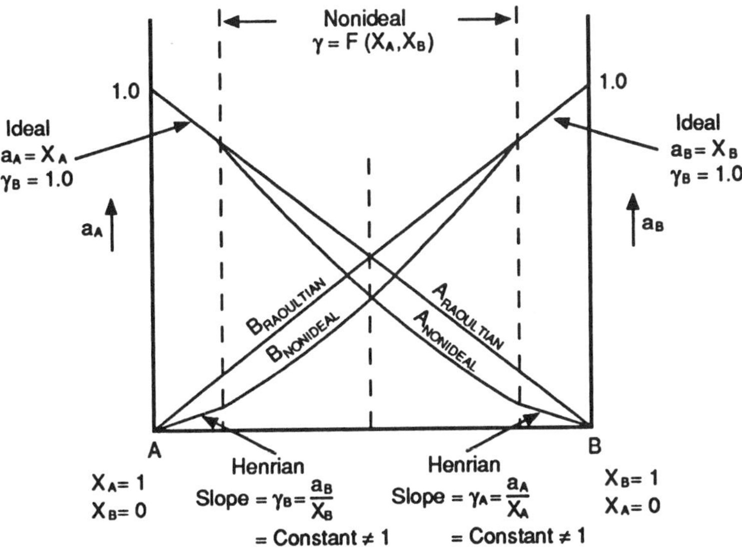

Figure 6.9 Nonideal binary solutions of A and B are represented by the line segments labeled $A_{Nonideal}$ and $B_{Nonideal}$. Solutes A and B exhibit Henrian behavior at low concentrations, while solvents A and B exhibit ideal behavior at high concentrations. At intermediate concentrations, A and B exhibit nonideal solution behavior.

6.8 CHEMICAL POTENTIAL

Recalling the four Maxwell equations for *closed systems* presented in Chapter Four, *total thermodynamic properties* for each equation can be expressed as follows:

$$U' = U'(S',V') \qquad [3\text{-}6]$$
$$H' = H'(S',P) \qquad [3\text{-}7]$$
$$G' = G'(T,P) \qquad [4\text{-}14]$$
$$A' = A'(T,V') \qquad [4\text{-}20]$$

Rewriting these equations for an *open* binary A-B system,

$$U' = U'(S',V',n_A,n_B)$$
$$H' = H'(S',P,n_A,n_B)$$
$$G' = G'(T,P,n_A,n_B)$$
$$A' = A'(T,V',n_A,n_B).$$

Using the chain rule to expand each expression for an *open system,*

$$dU' = \left(\frac{\partial U'}{\partial S'}\right)_{V',n_A,n_B} dS' + \left(\frac{\partial U'}{\partial V'}\right)_{S',n_A,n_B} dV' + \left(\frac{\partial U'}{\partial n_A}\right)_{S',V',n_B} dn_A + \left(\frac{\partial U'}{\partial n_B}\right)_{S',V',n_A} dn_B \quad [6\text{-}62]$$

$$dH' = \left(\frac{\partial H'}{\partial S'}\right)_{P,n_A,n_B} dS' + \left(\frac{\partial H'}{\partial P}\right)_{S',n_A,n_B} dP + \left(\frac{\partial H'}{\partial n_A}\right)_{S',P,n_B} dn_A + \left(\frac{\partial H'}{\partial n_B}\right)_{S',P,n_A} dn_B \quad [6\text{-}63]$$

$$dG' = \left(\frac{\partial G'}{\partial T}\right)_{P,n_A,n_B} dT + \left(\frac{\partial G'}{\partial P}\right)_{T,n_A,n_B} dP + \left(\frac{\partial G'}{\partial n_A}\right)_{T,P,n_B} dn_A + \left(\frac{\partial G'}{\partial n_B}\right)_{T,P,n_A} dn_B \quad [6\text{-}64]$$

$$dA' = \left(\frac{\partial A'}{\partial T}\right)_{V',n_A,n_B} dT + \left(\frac{\partial A'}{\partial V'}\right)_{T,n_A,n_B} dV' + \left(\frac{\partial A'}{\partial n_A}\right)_{T,V',n_B} dn_A + \left(\frac{\partial A'}{\partial n_B}\right)_{T,V',n_A} dn_B \quad [6\text{-}65]$$

By definition, μ_A, the *chemical potential* of component A is

$$\mu_A \equiv \left(\frac{\partial U'}{\partial n_A}\right)_{S',V',n_B} = \left(\frac{\partial H'}{\partial n_A}\right)_{S',P,n_B} = \left(\frac{\partial A'}{\partial n_A}\right)_{T,V',n_B} = \left(\frac{\partial G'}{\partial n_A}\right)_{T,P,n_B}$$

Since pressure and temperature are simplest to control during an experiment, the most useful of these expressions is

$$\mu_A = \left(\frac{\partial G'}{\partial n_A}\right)_{T,P,n_B} \quad [6\text{-}66]$$

Substituting [6-9] into [6-66],

$$\mu_A = \overline{G}_A \quad [6\text{-}67]$$

Equations [6-62] through [6-65] may be expressed in alternate form by incorporating the definitions of chemical potential into the extensive forms of [3-6], [3-7], [4-14], and [4-20]:

$$dU' = TdS' - PdV' + \mu_A dn_A + \mu_B dn_B \quad [6\text{-}68]$$
$$dH' = TdS' + V'dP + \mu_A dn_A + \mu_B dn_B \quad [6\text{-}69]$$
$$dG' = V'dP - S'dT + \mu_A dn_A + \mu_B dn_B \quad [6\text{-}70]$$
$$dA' = -S'dT - PdV' + \mu_A dn_A + \mu_B dn_B \quad [6\text{-}71]$$

Example Problem 6-9

Gases dissolve in metals monatomically and normally in dilute concentrations. Using TL analysis, derive *Sievert's law* relating the partial pressure of hydrogen in a gas phase to the concentration of hydrogen dissolved in a coexisting condensed phase at constant temperature. Assume the gas is ideal. Sievert's law is a special case of Henry's law (Kubaschewski and Alcock, 1979, p. 47).

Solution.
(1) Set Up.

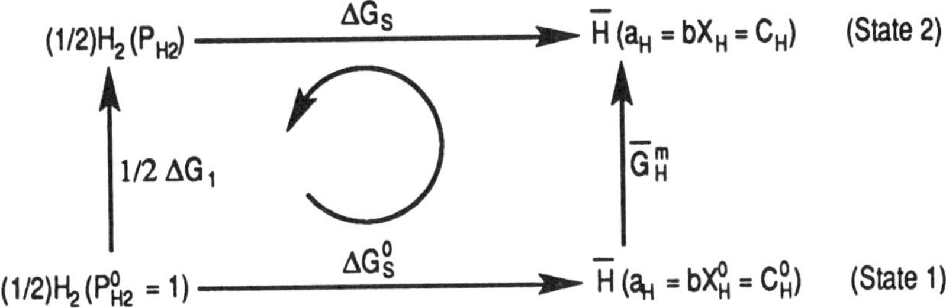

X_H^0 is the atomic fraction of $\overline{H}$ at $P_{H_2}^0 = 1$ atm. ΔG_s and ΔG_s^0 are the molar and standard molar Gibbs free energies of solution, respectively, relative to gaseous hydrogen.

(2) Sum.
$$\Sigma \Delta G_{TL} = 0 = \Delta G_s^0 + \overline{G}_H^m - \Delta G_s - \frac{1}{2}\Delta G_1.$$

(3) Substitute.
$$\Delta G_s^0 = \Delta H_s^0 - T\Delta S_s^0 .$$

$$\overline{G}_H^m = RT\ln(C_H / C_H^0).$$
$$\Delta G_s = 0 \text{ (equilibrium).}$$

$$\frac{1}{2}\Delta G_1 = (1/2)RT\ln(P_{H_2}).$$

Substituting into $\Sigma \Delta G_{TL} = 0$,

$$0 = \Delta H_s^0 - T\Delta S_s^0 + RT\ln(C_H / C_H^0) - (1/2)RT\ln(P_{H_2}).$$

(4) Solve.
$$RT\ln(C_H / C_H^0) = -\Delta H_s^0 + T\Delta S_s^0 + RT\ln(P_{H_2}^{1/2})$$

$$\ln[(C_H / C_H^0)/(P_{H_2}^{1/2})] = -\frac{\Delta H_s^0}{RT} + \frac{\Delta S_s^0}{R}$$

$$C_H = C_H^0\left[\exp\left(\frac{-\Delta H_s^0}{RT} + \frac{\Delta S_s^0}{R}\right)\right]P_{H_2}^{1/2} \qquad [6\text{-}72a]$$

$$C_H = kP_{H_2}^{1/2} \qquad [6\text{-}72b]$$

where C_H is the concentration of H in solution, C_H^0 is the concentration at $P_{H_2}^0 = 1$ atm, and ΔH_s^0 and ΔS_s^0 are the enthalpy and entropy of solution relative to gaseous hydrogen. In [6-72b], Sievert's law, C_H has arbitrary units consistent with constant k.

6.9 DISCUSSION QUESTIONS

(6.1) A component in a condensed solution is in equilibrium with its vapor. How is the Gibbs free energy change of the condensed component related to the vapor? What assumption is made about the vapor?

(6.2) At what relative solution concentrations (low, intermediate, or high) will a solution component exhibit:
(a) Ideal behavior?
(b) Dilute (Henrian) behavior?
(c) Regular behavior?

(6.3) Using a plot of H^m vs. X_A, show that $\overline{H}_A^m = 0$ for a solution ideal with respect to A over the entire composition from $X_A = 0$ to $X_A = 1.0$.

(6.4) Referring to (6.3), describe how to determine that component B is also ideal over the entire composition from $X_B = 0$ to $X_B = 1.0$.

(6.5) If $\overline{G}_i^m \to -\infty$ as $X_i \to 0$, what can be said about the stability of component i in solution as $X_i \to 0$? What does this imply about purification processes?

(6.6) Comment on the thermodynamic validity of the following expressions for a binary alloy:
(a) $\sigma = X_A \overline{\sigma}_A + X_B \overline{\sigma}_B$
where σ is stress and $\overline{\sigma}_A$ and $\overline{\sigma}_B$ are partial stresses.
What restriction is placed on σ?
(b) $C_p = X_A \overline{C}_p^A + X_B \overline{C}_p^B$
where C_p is constant pressure heat capacity and $\overline{C}_p^A$ and $\overline{C}_p^B$ are partial heat capacities at constant pressure.
(c) $\alpha = X_A \overline{\alpha}_A + X_B \overline{\alpha}_B$
where α is the thermal expansion coefficient and $\overline{\alpha}_A$ and $\overline{\alpha}_B$ are partial thermal expansion coefficients.

(6.7) The expression $\overline{G}_A^m$ only holds if the standard state is in the same state of aggregation as the solution. Explain.

(6.8) Excess Gibbs free energy of solution for a binary alloy is
$G^{XS} = X_A \overline{G}_A^{XS} + X_B \overline{G}_B^{XS}$.
Express G^{XS} for a solution that is subregular with respect to both A and B.

(6.9) Show that $\mu_A - \mu_A^0 = \overline{G}_A^m$.

(6.10) Is $\overline{H}_A = \mu_A$? Explain.

6.10 EXERCISE PROBLEMS

[6.1] Derive an expression equivalent to Sievert's law, [6-72b], for solution of hydrogen gas in an aqueous solution. Assume Henry's law.

Ans: $C_{H_2} = k\,P_{H_2}$.

[6.2] Sievert's constant for solution of oxygen in liquid silver is $k \approx 193.6$ cm^3·atm$^{-0.5}$/100 gm Ag at 1075°C (Darken and Gurry, 1953, p. 513). Calculate the solubility of Ag at the same temperature if $P_{O_2} = 50$ mm Hg.

Ans: $C_{Ag} = 49.7$ cm^3/100 gm Ag.

[6.3] Referring to Example Problem 6-5, show from the Gibbs-Duhem equation that $\overline{H}_{Au}^m = -20{,}083(1 - X_{Au})^2$ J/mol.

[6.4] Calculate the composition (a/o) of a binary A-B alloy in equilibrium with oxides A_2O and BO at 627°C. Given $\Delta G_{A_2O}^0 = -15{,}000$ cal/mol and $\Delta G_{BO}^0 = -25{,}000$ cal/mol. Assume solvent A is ideal. Solute B is Henrian with γ_B = constant = 1.15. The metallic phases are completely soluble in each other, whereas the oxides are insoluble.

Ans: 99.7 a/o A; 0.3 a/o B.

[6.5] For the binary liquid alloy system Cu-Zn, the zinc vapor pressure at 1060°C is given as a function of composition in the following partially completed table (Darken and Gurry, 1953, p. 512):

X_{Zn}	0.05	0.10	0.15	0.20	0.30	0.45	1.0
P_{Zn} (atm)	0.0289	0.0592	0.1184	0.2368	0.6000	1.2763	4.000*
a_{Zn}							
γ_{Zn}							
$\overline{G}_{Zn}^m$							
$\overline{G}_{Zn}^{m,XS}$							
$\overline{G}_{Zn}^{m,id}$							
Ω_{Zn}							

* Value is high compared to data in Appendix A, Table A.5.

(a) Complete the table (use metric units).

(b) Does the system obey the ideal, Henrian, or regular solution model?

Ans: Ideal behavior with respect to zinc as $X_{Zn} \to 1$, regular for approximately $0.05 < X_{Zn} < 0.30$, and Henrian for approximately $X_{Zn} < 0.05$.

[6.6] One mole of solid A at 1200°C is added to a large quantity of a liquid solution comprised of components A and B ($X_A = 0.8$). The liquid solution is also at 1200°C. If A and B form ideal solutions, calculate the Gibbs free energy change resulting from the addition of solid A to the mixture. Assume C_p^A (liquid) $\approx C_p^A$ (solid) and C_p^A is independent of temperature. $\Delta H_A^f = 24{,}000$ J/mol at 1800°C.

Ans: $\overline{G}_A^l - G_A^{o,s} = 4210$ J/mol.

[6.7] The molar heat of formation of liquid brass according to the reaction $(1 - X)Cu + XZn \rightarrow \overline{CuZn}$ is given by $H^m = -7100X(1 - X)$ cal/mol (Upadhyaya and Dube, 1977, p. 139). Determine the expressions relating the partial molar heats of mixing of copper and zinc in liquid brass to the alloy composition.

Ans: $\overline{H}_{Cu}^m = -7100(1 - X_{Cu})^2$; $\overline{H}_{Zn}^m = -7100(1 - X_{Zn})^2$.

[6.8] The solubility limit of component A in B is 1.5 w/o at 760°C. If an alloy contains 0.5 w/o A, calculate the activity of A. Assume Henrian solution behavior in the composition range.

Ans: $a_A = 0.33$.

[6.9] Antimony is removed from lead during the refining process by selective oxidation (Darken and Gurry, 1953, p. 513). Estimate the Sb content obtained for air agitation of the bath at 1173 K. State assumptions and use the following data: $\Delta G_{1100}^{0,f} = -417{,}600$ J/mol Sb_2O_3 and $\Delta G_{1200}^{0,f} = -395{,}000$ J/mol Sb_2O_3 (Wicks and Block, 1963, p. 13).

Ans: Applying two methods: Method 1 – antimony (ppb) ≈ 2 and Method 2 – (more realistic) Sb (ppm) ≈ 8000.

[6.10] Show for a regular binary solution that

$$\Omega_i = \frac{RT \ln(\gamma_i)}{(1 - X_i)^2}.$$

[6.11] Show that if a solution component i exhibits ideal behavior at constant temperature, $\overline{V}_i^{m,id} = 0$.

[6.12] During low temperature-high pressure metamorphism, the albite component in the mineral plagioclase breaks down to form the mineral quartz and jadeite, a component in the mineral omphacite. Quartz occurs as a separate phase during metamorphism, hence $a_{SiO_2} = 1$. Referring to Exercise Problem [4.22]:

(a) Calculate ΔG_T as a function of temperature, activity of albite, a_{Ab}, in plagioclase, and activity of jadeite, a_{Jd}, in omphacite.

Ans: ΔG_T (J/mol) $= 53.21 \times 10^3 - 2000.4T + 238.21T\ln(T) - 70.349 \times 10^{-3}T^2 + 3.791 \times 10^{-6}T^3 + 17.478 \times 10^3 T^{0.5} + 2460.4 \times 10^3 T^{-1} + RT \ln(a_{Jd}/a_{Ab})$.

(b) Assuming ideal mixing, determine ΔG_T as a function of temperature, X_{Ab}, and X_{Jd}.

Ans: ΔG_T (J/mol) $= 53.21 \times 10^3 - 2000.4T + 238.21T\ln(T) - 70.349 \times 10^{-3}T^2 + 3.791 \times 10^{-6}T^3 + 17.478 \times 10^3 T^{0.5} + 2460.4 \times 10^3 T^{-1} + RT \ln(X_{Jd}/X_{Ab})$.

[6.13] Plagioclase feldspar, one of the most common rock forming minerals, exhibits complete solid solution between the pure end members albite (Ab), $NaAlSi_3O_8$, and anorthite (An), $CaAl_2Si_2O_8$. Solid solution mineral compositions are usually given in abbreviated form, such as $Ab_{40}An_{60}$, which denotes a plagioclase of composition 40 w/o albite and 60 w/o anorthite. Show that the partial pressure of anorthite over a homogeneous liquid solution of $Ab_{20}An_{80}$ at 1350°C is:

$$P_{An} = \exp\left(\frac{-\Delta G^0_{1623} - 383R}{1623R}\right)$$

where ΔG^0_{1623} is the standard Gibbs free energy change at 1623 K for the phase transformation $An(l) \rightarrow An(g)$. The molecular weights of albite and anorthite are 262.23 and 278.21 respectively. State assumptions.

ANALYSIS AND APPLICATIONS: BINARY PHASE DIAGRAMS

7.1 EUTECTIC SYSTEMS: NO SOLID SOLUBILITY

Computation of thermodynamic properties of solutions from phase diagrams is now quite common although the complexities of the techniques described in the literature, particularly ternary systems, do not lend themselves easily to practical applications. The thermodynamic loop can be readily applied to phase diagram analysis. The TL is first superimposed directly around the equilibrium line of interest on the diagram (personal communication, 1959, R. Schuhmann, Jr., Department of Metallurgical Engineering, Purdue University, West Lafayette, Indiana). Various solution models are then tested to complete the analysis. In this chapter, the solution models used for analysis are ideal, regular, and dilute. Examples for eutectic systems with no solid solubility and terminal solubility are presented in this and subsequent sections. The chapter concludes with discussion of the Gibbs Phase rule.

As defined in Chapter 6, the parameters associated with $\overline{H}^m$ and $\overline{S}^m$ are independent of temperature. Temperature-composition data obtained directly from phase diagrams is correlated with thermodynamic expressions for liquidus, solidus, and solvus curves. Since fusion is incorporated into the analysis, a choice can be made whether or not to assume $\Delta C_p = 0$ for conversion of standard states from solid to liquid.

Consider the A-B alloy in Figure 7.1 for which the liquidus is thermodynamically characterized by the TL: $1 \rightarrow 2 \rightarrow 3 \rightarrow 4 \rightarrow 1$.

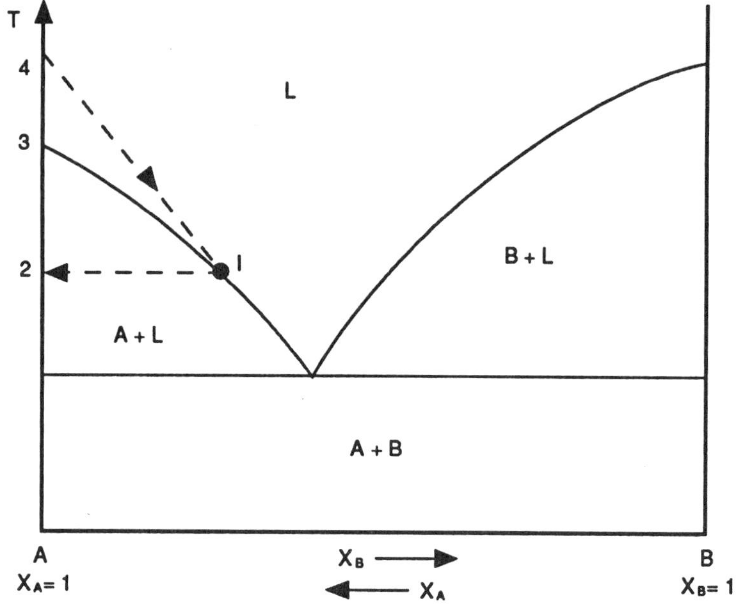

Figure 7.1 Temperature-composition eutectic phase diagram for components A and B with no solid solubility.

The following notation is used with reference to Figure 7.1:

Point	Phase Designation	Gibbs Free Energy
1	$\overline{A}\,(l)$ (Component A in liquid solution)	$\overline{G}_A^{\,l}$
2	$A^0(s)$ (Solid A)	$G_A^{0,s}$
3	$A^0(s) \rightarrow A^0(l)$ (Fusion of Solid A)	$G_A^{0,l} - G_A^{0,s}$
4	$A^0(l)$ (Liquid A)	$G_A^{0,l}$

Characterizing the thermodynamic loop:

Process	Reaction	Gibbs Free Energy Change
$1 \rightarrow 2$	$\overline{A}(l) \rightarrow A^0(s)$	$G_A^{0,s} - \overline{G}_A^l = 0.$ Solid A in equilibrium with A in liquid solution (from the diagram).
$2 \rightarrow 3 \rightarrow 4$	$A^0(s) \rightarrow A^0(l)$	$G_A^{0,l} - G_A^{0,s} = \Delta H_A^f +$ $\displaystyle\int_{T^f}^{T} \Delta C_p^A \, dT - T\left[\Delta S_A^f + \int_{T^f}^{T} (\Delta C_p^A / T)\,dT\right]$ where ΔH_A^f and ΔS_A^f are heat and entropy of fusion respectively and $\Delta C_p^A = C_p^{A(l)} - C_p^{A(s)}$.
$4 \rightarrow 1$	$A^0(l) \rightarrow \overline{A}(l)$	$\overline{G}_A^l - G_A^{0,l} = \overline{G}_A^{m,l} = \overline{H}_A^{m,l} - T\overline{S}_A^{m,l}.$

Summing about the TL in Figure 7.1,

$$\Sigma \Delta G_{\mathrm{TL}} = 0 = \Delta H_A^f + \int_{T^f}^{T} \Delta C_p^A \, dT - T\left[\Delta S_A^f + \int_{T^f}^{T} (\Delta C_p^A / T)\,dT\right] + \overline{H}_A^{m,l} - T\overline{S}_A^{m,l}.$$

Solving for T, the liquidus is given by

$$T = \frac{\overline{H}_A^{m,l} + \Delta H_A^f + \displaystyle\int_{T^f}^{T} \Delta C_p^A \, dT}{\overline{S}_A^{m,l} + \Delta S_A^f + \displaystyle\int_{T^f}^{T} \frac{\Delta C_p^A \, dT}{T}} \qquad [7\text{-}1]$$

For the specific case where $\Delta C_p^A \approx 0$,

$$T = \frac{\overline{H}_A^{m,l} + \Delta H_A^f}{\overline{S}_A^{m,l} + \Delta S_A^f} = \frac{\Sigma \Delta H}{\Sigma \Delta S} \qquad [7\text{-}2]$$

In general, temperature is equal to the sum of enthalpy terms divided by the sum of entropy terms. In order to relate [7-1] or [7-2] to the phase diagram, solution models are incorporated as follows:

Using [7-2]:
(1) Ideal Solution: substituting [6-27] and [6-28],

$$T = \frac{\Delta H_A^f}{-R \ln(X_A) + \Delta S_A^f} \qquad [7\text{-}3]$$

(2) Dilute Solution: substituting [6-35] and [6-36],

$$T = \frac{h + \Delta H_A^f}{s - R \ln(X_A) + \Delta S_A^f} \qquad [7\text{-}4]$$

(3) Regular Solution: substituting [6-39] and [6-42],

$$T = \frac{\Omega(1 - X_A)^2 + \Delta H_A^f}{-R \ln(X_A) + \Delta S_A^f} \qquad [7\text{-}5]$$

Example problems below illustrate the analysis of liquidus curves using trial and error combinations of [7-3], [7-4], and [7-5].

Example Problem 7-1

Analyze ideal and regular solution models for the Si liquidus of the Al-Si phase diagram. Plot the results for comparison on Figure E.6 in Appendix E.

Solution

From Table A.2, $\Delta H_{Si}^f = 50{,}630$ J/mol at $T_{Si}^f = 1693$ K.* From [4-8] at the melting point, $\Delta G_{1693}^f = 0 = \Delta H_{1693}^f - 1693\Delta S_{1693}^f$. Hence, $\Delta S_{1693}^f = 50{,}630/1693 = 29.91$ J/(mol·K).

Test (1): *Ideal Solution Model.* Applying [7-3],

$$T\,(\text{K}) = \frac{50{,}630}{-8.3144 \ln(X_{Si}) + 29.91}.$$

Selecting concentrations from $X_{Si} = 1.0$ to $X_{Si} = 0.122$ at the eutectic temperature, the liquidus temperature is calculated and tabulated in Table 7.1:

Table 7.1
Calculated Si Liquidus: Al-Si System—Ideal Solution Model
$T(\text{K}) = 50{,}630/[-8.3144\ln(X_{Si}) + 29.91]$

X_{Si}	T (K)	T (°C)
1.0	1693	1420
0.9	1645	1372
0.7	1540	1267
0.5	1419	1146
0.3	1268	995
0.122	1068	795

A plot of the data from Table 7.1 onto Figure E.6 reveals increasing deviation from the experimentally derived liquidus as X_{Si} decreases. As expected, the behavior of Si tends to be ideal at concentrations approaching $X_{Si}=1.0$.

---

* Use 1693 K for consistency with the equilibrium diagram.

Test (2): *Regular Solution Model.* Applying [7-5],

$$T\ (K) = \frac{\Omega^l (1 - X_{Si})^2 + 50,630}{-8.3144\ \ln(X_{Si}) + 29.91} \qquad [7\text{-}6]$$

The problem becomes one of finding a value for Ω^l that provides a reasonable fit to the liquidus curve. If such a value can be found, the solution tends to be regular. From [7-6],

$$\Omega^l = \{T[29.91 - 8.3144\ \ln(X_{Si})] - 50,630\}/(1 - X_{Si})^2.$$

Selecting concentrations, X_{Si}, and corresponding T from the liquidus, Ω^l is calculated and tabulated in Table 7.2. Data from the literature is also included.

Table 7.2

Calculated Ω^l and $\bar{H}^m_{Si}$ from the Si Liquidus: Al-Si System— Regular Solution Model

$$\Omega^l = \{T[29.91 - 8.3144\ln(X_{Si})] - 50,630\}/(1 - X_{Si})^2$$

X_{Si}	T(K)		Ω^l (J/mol) Calculated	$\bar{H}^m_{Si}$ (J/mol)= $\Omega^l_{Ave}\ (1\text{-}X_{Si})^2$ Calculated[a]
1.0	1687		$-\infty$	≈ 0[b]
0.9	1628		$-51,038$	≈ 0[b]
0.7	1513		-9881	-1104
				(-1406)
0.5	1333	$\Omega^l_{Ave.}=$	$-12,311$	-3067
				(-3598)
0.3	1103	$-12,267$	$-13,465$	-6011
				(-6276)
0.122	850		$-13,412$	-9456

(a) Non-parenthetical values are calculated using $\Omega^l_{Average} = -12,267$ J/mol. Parenthetical values are from Kubaschewski and Alcock, 1979, p. 389-90. The data from Kubaschewski and Alcock includes excess entropy whereas the regular solution does not. This may explain the difference between calculated and published values of $\bar{H}^m_{Si}$.

(b) At approximately $X_{Si} > 0.9$, $\bar{H}^m_{Si} \approx 0$ since the solution is virtually ideal with respect to Si(l) in this range. Note that the calculation for Ω^l is invalid at $X_{Si} = 1.0$.

Temperature is calculated as a function of composition from [7-6] using $\Omega^l_{Average}$ and tabulated in Table 7.3. A plot of the data from this table onto

Table 7.3
Calculated Si Liquidus: Al-Si System—Regular Solution Model
$$T(K) = [-12,267(1 - X_{Si})^2 + 50,630]/[-8.3144\ln(X_{Si}) + 29.91]$$

X_{Si}	T (K) (Calculated)	T (K) (Phase Diagram)
1.0	1693	1687
0.9	1641	1628
0.7	1506	1513
0.5	1333	1333
0.3	1118	1103
0.122	869	850

Figure E.6 reveals good correlation with the experimentally derived liquidus for a regular solution between the eutectic composition and $X_{Si} \approx 0.9$. Deviation from ideality is negative, therefore, mixing is exothermic. From [6-40], unlike pairs of atoms attract.

Example Problem 7-2

Calculate the equilibrium partial pressure of $N_2(g)$ over a eutectic liquid solution of Al-Si at the eutectic temperature. Assume $Si_3N_4(s)$ is a reaction product insoluble in the liquid solution. What is the limiting P_{N_2} below which $Si_3N_4(s)$ is reduced?

Solution
(1) Set Up.

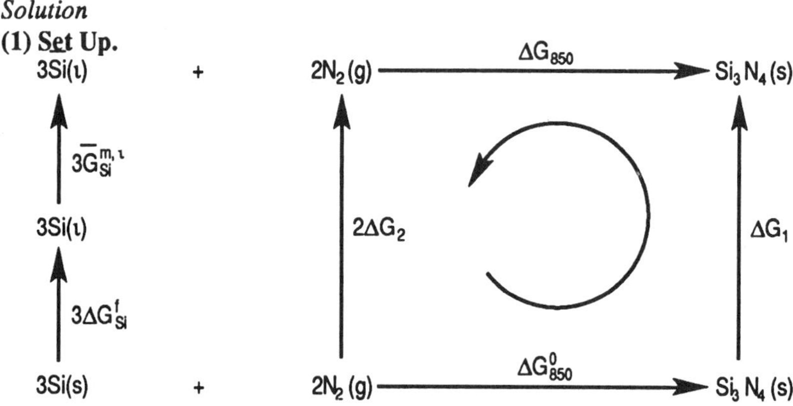

Note that the fusion of Si must be incorporated into the loop since Si is dissolved as a liquid in solution. In addition, $\overline{G}_{Si}^{m,l}$ is defined with respect to pure liquid Si according to [6-14].

(2) Sum.

$$\Sigma \Delta G_{TL} = 0 = \Delta G^0_{850} + \Delta G_1 - \Delta G_{850} - 2\Delta G_2 - 3\overline{G}^{m,l}_{Si} - 3\Delta G^f_{Si}.$$

(3) Substitute.

$\Delta G^0_{850} = -753{,}190 + 336.43T$ from Table A.4. From the Al-Si phase diagram, Figure E.6, the eutectic temperature $T_e = 850$ K, hence

$$\Delta G^0_{850} = -753{,}190 + 336.43(850) = -467{,}225 \text{ J/mol}.$$

$\Delta G_1 = 0.$

$\Delta G_{850} = 0$ (equilibrium).

$2\Delta G_2 = 2(8.3144)(850)\ln(P_{N_2}).$

Using the data in Example Problem 4-6 and assuming $\Delta C_p \approx 0$, the Gibbs free energy of fusion of Si(s) at 850 K is

$$3\Delta G^f_{Si} = 3[50{,}630 - 850(29.91)] = 75{,}620 \text{ J/mol}.$$

From Example Problem 7-1, $\Omega^l = -12{,}267$ J/mol and at the eutectic temperature $T_e = 850$ K, $X_{Si} = 0.122$. Substituting into [6-43],

$$3\overline{G}^{m,l}_{Si} = 3\left(\overline{H}^{m,l}_{Si} - T_e\overline{S}^{m,l}_{Si}\right) = 3\left\{\Omega^l(1 - X_{Si})^2 - T_e\left[-R\ln(X_{Si})\right]\right\}$$

$$= 3\left\{-12{,}267(1 - 0.122)^2 - 850[-8.3144\ln(0.122)]\right\}$$

$$= -72{,}972 \text{ J/mol}.$$

Substituting into $\Sigma \Delta G_{TL} = 0$,

$$0 = -467{,}225 - 2(8.3144)(850)\ln(P_{N_2}) - (-72{,}972) - 75{,}620.$$

(4) Solve.

$P_{N_2} = \underline{3.7 \times 10^{-15} \text{ atm}}$. This is the limiting P_{N_2} below which $Si_3N_4(s)$ tends to be reduced.

Example Problem 7-3

The solubility of iron in liquid lithium is 0.35 w/o at 1200°C and 0.004 w/o at 400°C. Given this information, (a) Estimate the maximum purity of liquid lithium with respect to iron that can be obtained by slowly cooling a high Li-Fe liquid solution. The eutectic temperature is only a fraction of a degree lower than the melting point of pure lithium; thus, the problem involves the calculation of the solubility of iron at the lowest temperature at which the solution is a liquid. For practical purposes, this temperature is the melting point of lithium. (b) Estimate the eutectic temperature and composition of the high Li-Fe liquid solution.

Solution

(a) At low concentration (< approximately 1 a/o solute) the solution is dilute with respect to iron solute, hence the solution model is characterized by [7-4]. A phase diagram is sketched in Figure 7.2 to illustrate the problem. Substituting ΔH_{Fe}^{f} and ΔS_{Fe}^{f} from Table A.2 into [7-4],

$$T = \frac{h_{Fe}^{l} + 13,770}{s_{Fe}^{l} - R\ln(X_{Fe}^{l}) + 7.6119} \qquad [7\text{-}7]$$

Since $X_{Fe}^{l} = (a/o)/100$, the solubility must be converted from w/o to atomic fraction. From [6-5a],

$$X_{Fe} = \frac{\text{w/o}|_{Fe}/55.85}{\left(\text{w/o}|_{Fe}/55.85\right) + \left(\text{w/o}|_{Li}/6.94\right)}.$$

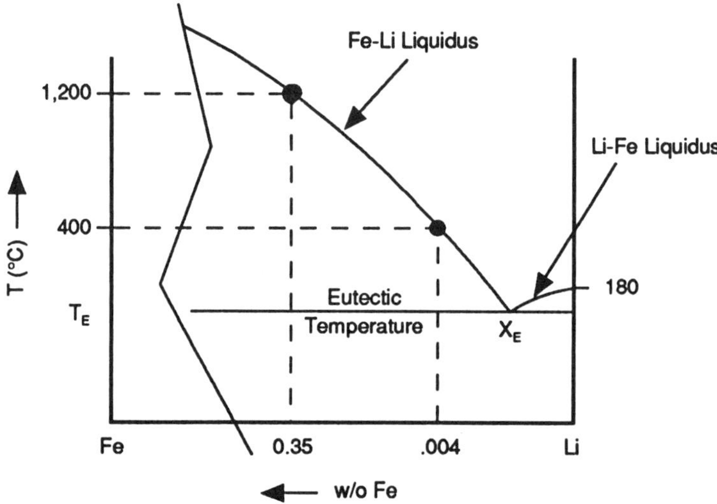

Figure 7.2 Partial temperature-composition phase diagram (not to scale): high lithium-iron eutectic, no solid solubility.

From the given data:
$X_{Fe} = 4.36 \times 10^{-4}$ at 0.35 w/o Fe and 1473 K;
$X_{Fe} = 4.97 \times 10^{-6}$ at 0.004 w/o Fe and 673 K.
Substituting and solving [7-7] simultaneously for two unknowns:

$$1473 = \frac{h_{Fe}^{l} + 13,770}{s_{Fe}^{l} - 8.3144\ln(4.36\times10^{-4}) + 7.6119};$$

$$673 = \frac{h_{Fe}^{l} + 13,770}{s_{Fe}^{l} - 8.3144\ln(4.97\times10^{-6}) + 7.6119}$$

hence, $\qquad\qquad h_{Fe}^{l} = 32,331$ J/mol;

$$s_{Fe}^{l} = -40.65 \text{ J/(mol·K)}.$$

136

Inserting these values into [7-7] and collecting terms:

$$T = \frac{46,101}{-33.038 - 8.3144\,\ln(X_{Fe}^{l})} \qquad [7\text{-}8]$$

The minimum iron content in lithium at the melting point of lithium (453 K) is estimated to be

$$\ln(X_{Fe}^{l}) = \frac{46,101 + 33.038T}{-8.3144T}$$

$$= \frac{46,101 + 33.038(453)}{-8.3144(453)}$$

or

From [6-5b], $\qquad \underline{X_{Fe}^{l} = 9.089 \times 10^{-8} \approx 0.7\ \text{ppm}}$.

(b) In Example Problem 6-8, it was shown if the solute is dilute, the solvent is ideal. Substituting $\Delta H_{Li}^{f} = 2929$ J/mol and $\Delta T_{Li}^{f} = 453$ K into [7-3],

$$T = \frac{2929}{-8.3144\,\ln(X_{Li}) + 2929/453} \qquad [7\text{-}9]$$

Solving [7-8] and [7-9] simultaneously for T_e (eutectic temperature) and X_e (eutectic composition):

$$T_e = \frac{46,101}{-33.038 - 8.3144\,\ln\!\left(X_e\right)};$$

$$T_e = \frac{2929}{-8.3144\,\ln\!\left(1 - X_e\right) + 6.466}$$

where

$$X_e = X_{Fe}\ \text{and}\ 1 - X_e = X_{Li}$$

hence, $\qquad \underline{X_e = 9.083 \times 10^{-8}};$

$$\underline{T_e = 452.9\,\text{K}\,(179.9°\text{C})}.$$

Example Problem 7-4

The solubility of carbon in liquid aluminum is 6 ppm at 960°C and 12.5 ppm at 1000°C.* Predict the solubility at the melting point of Al, 660°C. Assume dilute behavior.

Solution

Since carbon as graphite does not have a known fusion temperature, use pure

---

* According to Simensen (1989, p. 191), Al-C melts are saturated with carbon and also contain traces of carbides (Al$_4$C$_3$).

solid C as the standard state. As a result, fusion terms in [7-4], are dropped and

$$T = \frac{h}{s - R\ln(X_C)} \qquad [7\text{-}10]$$

Converting solubility data to atomic fraction,

At 960°C:
$$X_C = \frac{6/12}{6/12 + \approx 10^6/26.984}$$
$$= 13.49 \times 10^{-6} \approx 6 \text{ ppm.}$$

At 1000°C:
$$X_C = \frac{12.5/12}{12.5/12 + \approx 10^6/26.984}$$
$$= 28.11 \times 10^{-6} \approx 12.5 \text{ ppm.}$$

Substituting into [7-10],
$$1233 = h_C/[s - 8.3144\ln(13.49 \times 10^{-6})];$$
$$1273 = h_C/[s - 8.3144\ln(28.11 \times 10^{-6})].$$

Solving simultaneously,
$$h_C = 239{,}490 \text{ J/mol;}$$
$$s_C = 101 \text{ J/(mol·K).}$$

Substituting h_C, s_C, and $T = 660$°C (993 K) into [7-10] and solving for X_C,

From [6-5b], $\qquad X_C = 7.38 \times 10^{-9} \approx 3 \text{ ppb .}$

7.2 EUTECTIC SYSTEMS: TERMINAL SOLID SOLUBILITY

Consider the A-B alloy in Figure 7.3 for which the liquidus and solidus are thermodynamically characterized by the TL: $1 \rightarrow 2 \rightarrow 3 \rightarrow 4 \rightarrow 5 \rightarrow 1$.

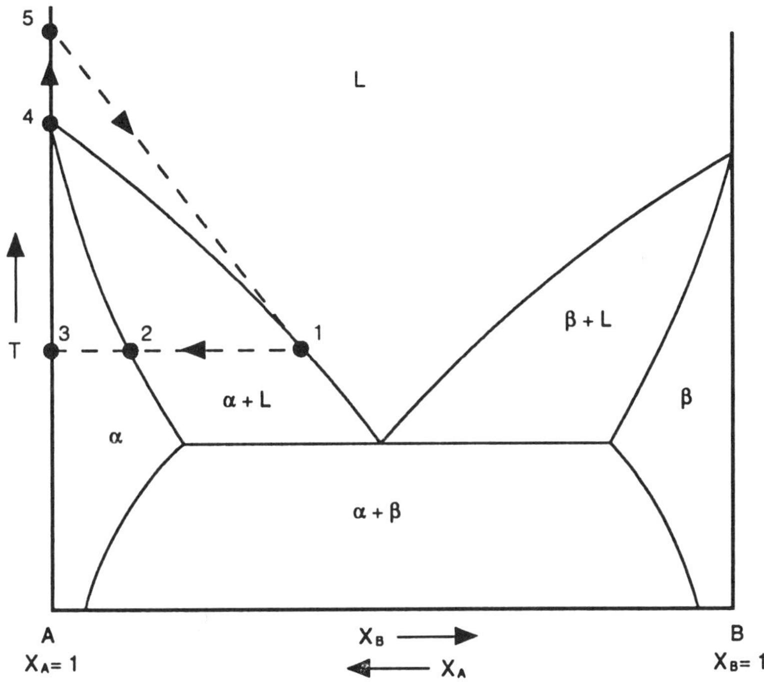

Figure 7.3 Temperature-composition eutectic phase diagram for components A and B with terminal solid solubility.

The following notation is used with reference to Figure 7.3:

Point	Phase Designation	Gibbs Free Energy
1	$\overline{A}(l)$ (Component A in liquid solution)	$\overline{G}_A^l$
2	$\overline{A}(\alpha)$ (Component A in α solid solution)	$\overline{G}_A^\alpha$
3	$A^0(s)$ (Solid A)	$G_A^{0,s}$
4	$A^0(s) \rightarrow A^0(l)$ (Fusion of Solid A)	$G_A^{0,l} - G_A^{0,s}$
5	$A^0(l)$ (Liquid A)	$G_A^{0,l}$

Characterizing the thermodynamic loop:

Process	Reaction	Gibbs Free Energy Change
$1 \rightarrow 2$	$\overline{A}(l) \rightarrow \overline{A}(\alpha)$	$\overline{G}_A^\alpha - \overline{G}_A^l = 0$
$2 \rightarrow 3$	$\overline{A}(\alpha) \rightarrow A^\circ(s)$	$G_A^{0,s} - \overline{G}_A^\alpha = -\overline{G}_A^{m,\alpha}$
		where $-\overline{G}_A^{m,\alpha} = -\overline{H}_A^{m,\alpha} + T\overline{S}_A^{m,\alpha}$
$3 \rightarrow 4 \rightarrow 5$	$A^0(s) \rightarrow A^0(l)$	$G_A^{0,l} - G_A^{0,s} = \Delta H_A^f - T\Delta S_A^f$
$5 \rightarrow 1$	$A^0(l) \rightarrow \overline{A}(l)$	$\overline{G}_A^l - G_A^{0,l} = \overline{G}_A^{m,l}$
		where $\overline{G}_A^{m,l} = \overline{H}_A^{m,l} - T\overline{S}_A^{m,l}$

Summing about the TL in Figure 7.3,

$$\Sigma \Delta G_{TL} = 0 = 0 - \overline{H}_A^{m,\alpha} + T\overline{S}_A^{m,\alpha} + \Delta H_A^f - T\Delta S_A^f + \overline{H}_A^{m,l} - T\overline{S}_A^{m,l}.$$

Solving for T and assuming $\Delta C_p^A = C_p^{A(l)} - C_p^{A(\alpha)} \approx 0$, the equation for the liquidus* is:

$$T = \frac{\overline{H}_A^{m,l} - \overline{H}_A^{m,\alpha} + \Delta H_A^f}{\overline{S}_A^{m,l} - \overline{S}_A^{m,\alpha} + \Delta S_A^f} \qquad [7\text{-}11]$$

As before, expressions for specific solution models are substituted into [7-11], resulting in the following liquidus equations:

(1) **Ideal Solution:** substituting [6-27] and [6-28],

$$T = \frac{\Delta H_A^f}{-R\ln(X_A^l) + R\ln(X_A^\alpha) + \Delta S_A^f} \qquad [7\text{-}12]$$

(2) **Regular Solution:** substituting [6-39] and [6-42],

$$T = \frac{\Omega^l(1 - X_A^l)^2 - \Omega^\alpha(1 - X_A^\alpha)^2 + \Delta H_A^f}{-R\ln(X_A^l) + R\ln(X_A^\alpha) + \Delta S_A^f} \qquad [7\text{-}13]$$

Example Problem 7-5

Assuming Ag-Cu solid and liquid solutions are regular, calculate the parameters Ω^l and Ω^α from analysis of the Ag-Cu phase diagram in Appendix E, Figure E.7.

Solution

From Figure E.7, two sets of data points are tabulated as follows:

$$T = 1201 \text{ K}, \ X_{Cu}^l \approx 0.78, \ X_{Cu}^\alpha \approx 0.96;$$

---

* An expression similar to [7-1] can be derived assuming $\Delta C_p \neq 0$.

$$T = 1052 \text{ K}, \ X_{Cu}^{l} \approx 0.40, \ X_{Cu}^{\alpha} \approx 0.94.$$

Substituting these values and $\Delta H_{Cu}^{f} = 12{,}972$ J/mol from Table A.2 into [7-13], simultaneous equations are solved for Ω^{l} and Ω^{α}:

$$1201 = \frac{\Omega^{l}(1-0.78)^2 - \Omega^{\alpha}(1-0.96)^2 + 12{,}972}{8.3144 \ln(0.96/0.78) + 9.566};$$

$$1052 = \frac{\Omega^{l}(1-0.40)^2 - \Omega^{\alpha}(1-0.94)^2 + 12{,}972}{8.3144 \ln(0.94/0.40) + 9.566}.$$

Hence, $\quad \Omega^{l} = \underline{12{,}889}$ J/mol, $\overline{H}_{Cu}^{m,l} = 12{,}889(1 - X_{Cu}^{l})^2;$

$$\Omega^{\alpha} = \underline{21{,}111} \text{ J/mol}, \ \overline{H}_{Cu}^{m,\alpha} = 21{,}111(1 - X_{Cu}^{\alpha})^2.$$

As mentioned in Section 6.4, the regular solution model is ideal with respect to entropy, hence from Section 6.5, $\overline{S}_{Cu}^{XS} = 0$. Ω is independent of temperature and composition. A comparison of $\overline{H}_{Cu}^{m,l}$ computed above with data from the literature is given in Table 7.4. Considering the errors inherent in parameters calculated from phase diagrams (Kubaschewski and Alcock, 1979, p. 50-52) and the assumptions listed above, the results are comparable. In addition, the same authors report an excess entropy contribution which suggests that the solution is not strictly regular.

Table 7.4

Calculated $\overline{H}_{Cu}^{m,l}$ Versus Published $\overline{H}_{Cu}^{m,l}$

Ag-Cu Phase Diagram

Source*		$X_{Cu}^{l} = 0.5$	$X_{Cu}^{l} = 0.7$	$X_{Cu}^{l} = 0.9$
K & A: (1423 K)	$\overline{H}_{Cu}^{m,l}$ (J/mol)	3766	1435	159
Calculated: (1050–1200 K)	$\overline{H}_{Cu}^{m,l}$ (J/mol)	3222	1160	129

* K & A: Kubaschewski and Alcock, 1979, p. 387.

Example Problem 7-6

During *gas-carburizing*, steel parts are placed in a furnace with an atmosphere containing hydrocarbon gases such as methane (CH_4) (Smith 1986, p. 153-158). Carbon diffuses into the surface of the steel and subsequent heat treatment results in a product with a wear-resistant high-carbon case.

Suppose a low carbon steel is carburized at 900°C in a hydrogen-methane mixture. The surface carbon content required is eutectoid in composition or approximately 0.8 w/o.

(a) Calculate the *carburizing potential* $P_{CH_4}/P_{H_2}^2$ required to develop the necessary surface carbon. Note: $\overline{C}(s)$ is carbon dissolved in austenite.

Solution
(1) Set Up.

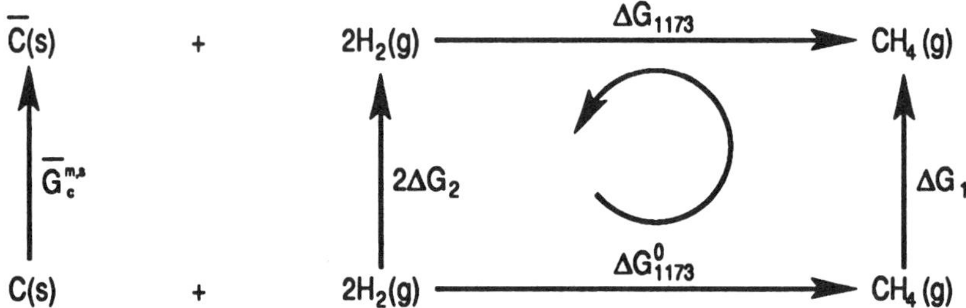

(2) Sum.

$$\Sigma \Delta G_{TL} = 0 = \Delta G_{1173}^0 + \Delta G_1 - \Delta G_{1173} - 2\Delta G_2 - \overline{G}_C^{m,s}.$$

(3) Substitute. From Table A.4,

$$\Delta G_{1173}^0 = -69{,}126 + 51.26T \log(T) - 65.36T$$
$$= -69{,}126 + 51.26(1173)\log(1173) - 65.36(1173)$$
$$= 38{,}757 \text{ J/mol.}$$

$\Delta G_1 = RT \ln(P_{CH_4})$ and $2\Delta G_2 = 2RT \ln(P_{H_2}) = RT \ln(P_{H_2}^2)$.

$\Delta G_{1173} = 0$ (equilibrium).

$\overline{G}_C^{m,s}$ is obtained from Appendix A, Table A.6. The activity of carbon relative to graphite is expressed by the constant temperature conversion a_c (graphite) = a_c (w/o C in austenite)/a_c (w/o C in austenite at saturation). Hence, a_c (0.80 w/o in steel)= 0.048 at 900°C. From the Fe-C phase diagram, the carbon content of saturated austenite at 900°C is 1.18 w/o C. By interpolation between 1.1 and 1.2 w/o C, a_c = 0.0803 at 1.18 w/o C.

From [6-14], $\overline{G}_C^{m,s}$= 8.3144(1173)ln(0.0480/0.0803). Substituting the above data into $\Sigma \Delta G_{TL} = 0$, $0 = 38{,}757 + 8.3144(1173) \ln(P_{CH_4}) - 8.3144(1173) \ln(P_{H_2}^2) - 8.3144(1173)\ln(0.0480/0.0803)$.

(4) Solve.

$$P_{CH_4}/P_{H_2}^2 = \underline{0.011}.$$

(b) Assuming $P_{CH_4} + P_{H_2} = 1$ atm, calculate the partial pressure of each gas.

Solution

Solving simultaneously:

From (a) $P_{CH_4} / P_{H_2}^2 = 0.011$;

From (b) $P_{CH_4} + P_{H_2} = 1$ atm.

Hence,

$$P_{H_2}^2 + 90.909\, P_{H_2} - 90.909 = 0,$$

$$P_{H_2} = \{-90.909 + [(90.909)^2 - 4(-90.909)]^{1/2}\}/2 \quad \text{or}$$

$$P_{H_2} = \underline{0.989\ \text{atm}}; \quad P_{CH_4} = 1 - 0.989 = \underline{0.011\ \text{atm}}.$$

Note that excess CH_4 ($P_{CH_4} > 0.011$ atm) drives the reaction from right to left, hence CH_4 is a *carburizing gas*. Conversely, excess H_2 ($P_{H_2} > 0.989$ atm) drives the reaction from left to right, hence H_2 is a *decarburizing gas*.

7.3 CHEMICAL POTENTIAL: PHASES AT EQUILIBRIUM

Referring to Figure 7.4, the ends of the tie-line through point X connect co-existing terminal solid solution phases α and β. The total Gibbs free energy change of the *system* can be found by substituting [4-25], [4-26], and [6-66] into [6-64]:

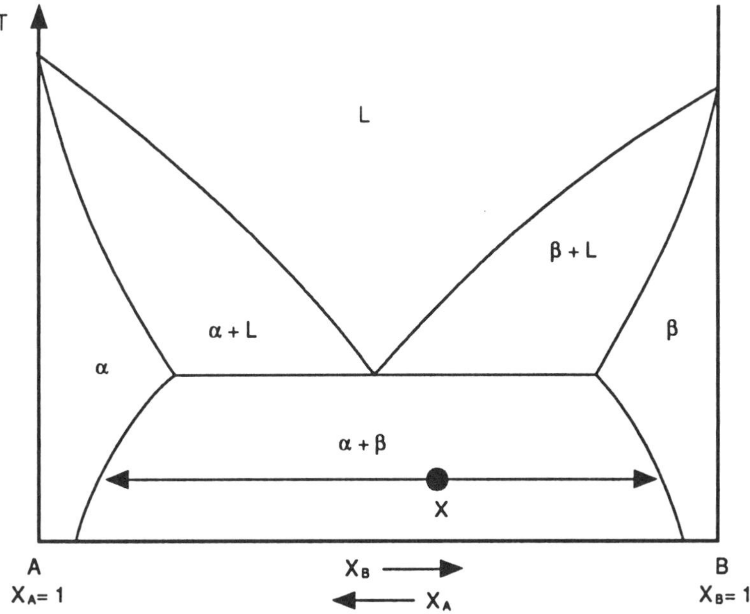

Figure 7.4 Temperature-composition eutectic phase diagram for components A and B with terminal solid solubility. Point X is a composition in the two phase $\alpha + \beta$ field.

$$dG' = dG'_\alpha + dG'_\beta$$
$$= -S'_\alpha dT + V'_\alpha dP + \mu_A^\alpha dn_A^\alpha + \mu_B^\alpha dn_B^\alpha - S'_\beta dT + V'_\beta dP + \mu_A^\beta dn_A^\beta + \mu_B^\beta dn_B^\beta$$

where μ_A^α and dn_A^α are the *chemical potential* of component A in α phase and the differential of the number of moles of A in α respectively. Definitions with respect to β phase are identical. If α and β are in equilibrium at constant temperature and pressure, $dG' = 0$, hence:

$$\mu_A^\alpha dn_A^\alpha + \mu_B^\alpha dn_B^\alpha + \mu_A^\beta dn_A^\beta + \mu_B^\beta dn_B^\beta = 0 \qquad [7\text{-}14]$$

From a mass balance on A,

$$n_A^\alpha + n_A^\beta = n_A .$$

Since the total number of moles of A is constant, $dn_A^\alpha + dn_A^\beta = dn_A = 0$. Hence, for equilibrium to exist between phases α and β,

$$dn_A^\alpha = -dn_A^\beta .$$

Similarly, $\qquad\qquad\qquad dn_B^\alpha = -dn_B^\beta .$

Substituting these equalities into [7-14],

$$\mu_A^\alpha(-dn_A^\beta) + \mu_B^\alpha(-dn_B^\beta) + \mu_A^\beta dn_A^\beta + \mu_B^\beta dn_B^\beta = 0$$

or

$$(\mu_A^\alpha - \mu_A^\beta)dn_A^\beta + (\mu_B^\alpha - \mu_B^\beta)dn_B^\beta = 0.$$

Regardless of the change in the number of moles in each phase, if equilibrium is to be satisfied,

$$\mu_A^\alpha = \mu_A^\beta \qquad [7\text{-}15]$$

$$\mu_B^\alpha = \mu_B^\beta \qquad [7\text{-}16]$$

[7-15] and [7-16] may be extended to any number of phases (solid, liquid, or gas) at equilibrium.

Example Problem 7-7

Determine the activity coefficient, γ_{Cu}^β, in a high Ag-Cu alloy at 500°C.

Solution

Examination of the Ag-Cu alloy phase diagram in Figure E.7 reveals that equilibrium involves two solvus transformations, α and β, at 500°C. At equilibrium, $\mu_{Cu}^\alpha = \mu_{Cu}^\beta$. Assume: (1) Cu behaves ideally in α phase and Henrian in β phase and (2) the standard state for Cu is pure solid copper. Substituting [6-67] into [7-15], $\overline{G}_{Cu}^\alpha = \overline{G}_{Cu}^\beta$. Subtracting G_{Cu}^0, $\overline{G}_{Cu}^\alpha - G_{Cu}^0 = \overline{G}_{Cu}^\beta - G_{Cu}^0$. Substituting G for V in [6-7]:

$$\overline{G}_{Cu}^{m,\alpha} = \overline{G}_{Cu}^{m,\beta} .$$

From [6-14], $\qquad\qquad RT\ln(a_{Cu}^\alpha) = RT\ln(a_{Cu}^\beta),$

$$a_{Cu}^{\alpha} = a_{Cu}^{\beta};$$

$$X_{Cu}^{\alpha} = \gamma_{Cu}^{\beta} X_{Cu}^{\beta}.$$

From the phase diagram at 500°C, the terminal solubility of Ag in α is $\approx$ 2 w/o and the terminal solubility of Ag in β is $\approx$ 98 w/o. Converting to a/o:

2 w/o Ag = 1.19 a/o Ag = 98.81 a/o Cu in α;

98 w/o Ag = 95.65 a/o Ag = 3.35 a/o Cu in β.

Hence, $$X_{Cu}^{\alpha} = \gamma_{Cu}^{\beta} X_{Cu}^{\beta} \Rightarrow 0.9881 = \gamma_{Cu}^{\beta}(0.0335)$$

or $$\gamma_{Cu}^{\beta} = \underline{29.5}.$$

As in previous problems, oxidation, sulfidizing, or chloridizing *potential* (gas composition ratios) can be determined for alloyed components if activities in the alloy are obtained from the literature or calculated by first determining γ as described above.

Example Problem 7-8

Develop an expression relating temperature to composition in the two phase $\alpha + \beta$ field of the Ag-Cu system.

Solution

At a temperature slightly below the eutectic, solid compositions are defined by α and β solvus boundaries as shown in Figure 7.5.

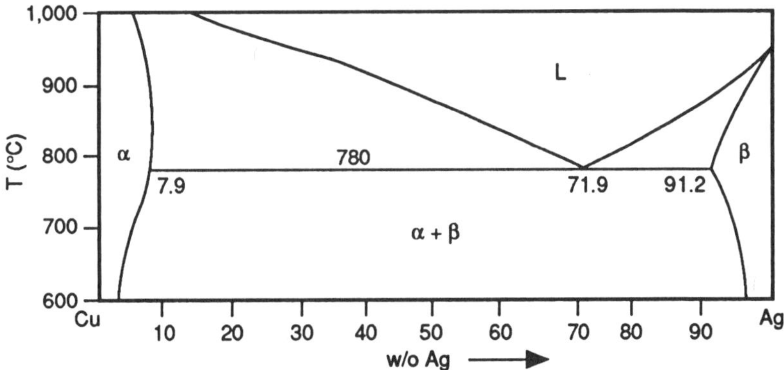

Figure 7.5 Partial Ag-Cu phase diagram at and below the eutectic temperature.

At equilibrium in the two phase field, assuming a standard state of pure solid copper,

$$\overline{G}_{Cu}^{m,\alpha} = \overline{G}_{Cu}^{m,\beta} \text{ where } \overline{G}_{Cu}^{\alpha} = \mu_{Cu}^{\alpha} \text{ and } \overline{G}_{Cu}^{\beta} = \mu_{Cu}^{\beta}$$

Hence,

$$\overline{H}_{Cu}^{m,\alpha} - T\overline{S}_{Cu}^{m,\alpha} = \overline{H}_{Cu}^{m,\beta} - T\overline{S}_{Cu}^{m,\beta} \ .$$

Solving for T,

$$T = \frac{\overline{H}_{Cu}^{m,\alpha} - \overline{H}_{Cu}^{m,\beta}}{\overline{S}_{Cu}^{m,\alpha} - \overline{S}_{Cu}^{m,\beta}} \qquad [7\text{-}17]$$

In Example Problem 7-7, it was assumed that Cu is ideal in α phase and Henrian in β phase at 500°C. If it is assumed, rather, that both solutions are regular with respect to Cu at higher concentrations near 780°C, [7-17] becomes

$$T = \frac{\Omega^{\alpha}(1 - X_{Cu}^{\alpha})^2 - \Omega^{\beta}(1 - X_{Cu}^{\beta})^2}{R \ln(X_{Cu}^{\beta} / X_{Cu}^{\alpha})} \ .$$

Since $\Omega^{\alpha} = 21{,}111$ J/mol from Example Problem 7-5, the solution becomes one of finding Ω^{β} and substituting it back into the above equation. At the eutectic temperature,

$$1053 = \frac{21{,}111(1 - 0.952)^2 - \Omega^{\beta}(1 - 0.141)^2}{8.3144 \ln(0.141 / 0.952)} \ .$$

Hence, $\Omega^{\beta} = 22{,}720$ J/mol. The expression for T becomes:

$$T = \frac{21{,}111(1 - X_{Cu}^{\alpha})^2 - 22{,}720(1 - X_{Cu}^{\beta})^2}{R \ln(X_{Cu}^{\beta} / X_{Cu}^{\alpha})} \ .$$

An alternate method of finding Ω^{β} would be analysis of the Ag liquidus/solidus as is done for the Cu liquidus/solidus in Example Problem 7-5.

7.4 UNIVARIANT EQUILIBRIUM: CLAPEYRON EQUATION

The functional dependence between pressure and temperature for univariant equilibrium can be expressed in terms of the *Clapeyron equation* derived below. For example, consider the *liquid-vapor* transformation for pure component A:

$$A(l) \rightarrow A(g) \qquad \Delta G = G_{A(g)} - G_{A(l)}.$$

From [4-14],

$$dG_{A(l)} = V_{A(l)}dP - S_{A(l)}dT;$$
$$dG_{A(g)} = V_{A(g)}dP - S_{A(g)}dT.$$

At equilibrium, $\quad dG_{A(l)} = dG_{A(g)}\quad$ hence,

$$V_{A(l)}dP - S_{A(l)}dT = V_{A(g)}dP - S_{A(g)}dT;$$
$$[S_{A(g)} - S_{A(l)}]dT = [V_{A(g)} - V_{A(l)}]dP$$

or

$$dP = \left(\frac{S_{A(g)} - S_{A(l)}}{V_{A(g)} - V_{A(l)}} \right) dT \qquad [7\text{-}18]$$

Since the reaction proceeds at constant temperature and pressure, [3-9] substituted into [7-18] gives

$$dP = \frac{\left[H_{A(g)} - H_{A(l)}\right]dT}{\left[V_{A(g)} - V_{A(l)}\right]T} \approx \frac{\left[H_{A(g)} - H_{A(l)}\right]dT}{TV_{A(g)}} \qquad [7\text{-}19]$$

where $V_{A(g)} \gg V_{A(l)}$; $V_{A(g)} - V_{A(l)} \approx V_{A(g)}$. [7-19] is one form of the *Clapeyron equation*. Assuming ideal gas behavior, [1-1] substituted into [7-19] results in an alternate form of the Clapeyron equation sometimes referred to as the *Clausius-Clapeyron equation*:

$$\frac{dP}{P} = \frac{\left[H_{A(g)} - H_{A(l)}\right]dT}{RT^2}$$

or

$$\frac{dP}{P} = \frac{\Delta H^V dT}{RT^2} \qquad [7\text{-}20]$$

where ΔH^v is the molar heat of vaporization. For a *solid-vapor* transformation, "*l*" for liquid in [7-18] and [7-19] is replaced by "*s*" for solid, and ΔH^v in [7-20] is replaced by ΔH^s for molar heat of sublimation.

For a *liquid-solid* or *solid-solid* (α-β) transformation,

$$A(\alpha) \rightarrow A(\beta);$$

$$dP = \frac{\left[H_{A(\beta)} - H_{A(\alpha)}\right]dT}{\left[V_{A(\beta)} - V_{A(\alpha)}\right]T} = \frac{\Delta H^{Tr} dT}{T\Delta V} \qquad [7\text{-}21]$$

where ΔH^{Tr} and ΔV denote the molar enthalpy and volume of transformation respectively. ΔH^{Tr} and ΔV may be assumed constant over small temperature intervals to simplify integration of [7-21].

7.5 PRESSURE EFFECT ON PHASE BOUNDARIES

Consider a simple eutectic system with no solid solubility. The effect of pressure on the liquidus will now be estimated. From [4-32] at constant temperature,

$$dH = V(1 - \alpha T)dP \qquad [7\text{-}22]$$

Substituting [1-2] into [4-23],

$$dS = -\alpha V dP \qquad [7\text{-}23]$$

where

$$(\partial V / \partial T)_P = \alpha V.$$

Since α is a thermodynamic property, $\overline{\alpha}_A^{m,l}$ is defined as a partial molar mixing property. Substituting the pressure correction terms $\alpha_A^{o,l}$, $V_A^{o,l}$, $\overline{\alpha}_A^{m,l}$, $\overline{V}_A^{m,l}$, [7-22], and [7-23] into [7-2] ,

$$T = \frac{\bar{H}_A^{m,l} + \displaystyle\int_{P_1}^{P_2} \bar{V}_A^{m,l}(1 - \bar{\alpha}_A^{m,l}T)dP + \Delta H_A^f + \displaystyle\int_{P_1}^{P_2} V_A^{0,l}(1 - \alpha_A^{0,l}T)dP}{\bar{S}_A^{m,l} - \displaystyle\int_{P_1}^{P_2} \bar{V}_A^{m,l}\bar{\alpha}_A^{m,l}dP + \Delta S_A^f - \displaystyle\int_{P_1}^{P_2} V_A^{0,l}\alpha_A^{0,l}dP} \qquad [7\text{-}24a]$$

The above equation is simplified in the numerator by neglecting αT terms since $\bar{\alpha}_A^{m,l}T \approx \alpha_A^{0,l}T \ll 1$. In the denominator, $\bar{V}_A^{m,l}\bar{\alpha}_A^{m,l} \approx V_A^{0,l}\alpha_A^{0,l} \ll (\bar{S}_A^{m,l} + \Delta S_A^f)$. Assuming $\bar{V}_A^l \approx V_A^{0,l}$ and $V_A^{0,l}$ is independent of pressure, the above equation reduces to

$$T = \frac{\bar{H}_A^{m,l} + V_A^{0,l}(P_2 - P_1) + \Delta H_A^f}{\bar{S}_A^m + \Delta S_A^f} \qquad [7\text{-}24b]$$

Example Problem 7-9

Predict the liquidus temperature shift caused by increasing the hydrostatic pressure from 1 to 1000 atm over a 60 a/o Si-Al alloy.

Solution

From Example Problem 7-1, $\Delta H_{Si}^f = 50{,}630$ J/mol, $\bar{H}_{Si}^{m,l} = -12{,}267(1 - X_{Si})^2$ and $\Delta S_{Si}^f = 50{,}630/1693 = 29.91$ J/(mol·K). From Table B.1, $V_{Si}^{0,l} = M_{Si}/\rho_{Si}$ $= 28.09/2.57 = 10.93$ cm^3/mol. Substituting into [7-24b],

$$T = \frac{-12{,}267(1 - 0.6)^2 + (10.93)(999)(0.101) + 50{,}630}{-8.3144\ln(0.6) + 29.91}$$

$$= \frac{49{,}770}{34.16} = 1457 \text{ K.}$$

Since the liquidus temperature is ≈ 1424 K at $X_{Si} = 0.6$ and 1 atm, the liquidus shifts upward by <u>approximately 33 K</u>. In a high Al-Si solution, Wu (1992, p. 1-5) predicted an upward shift of ≈ 50 K for an increase in pressure from 1 to 6800 atm. The above assumptions become less valid at higher pressures.

Example Problem 7-10

Develop an expression that predicts the effect of pressure on solvus boundaries in the $\alpha + \beta$ region of Figure 7.3.

Solution

Starting with [7-17], incorporate pressure correction terms corresponding to those used to develop [7-24a]:

$$T = \frac{\bar{H}_A^{m,\alpha} + \displaystyle\int_{P_1}^{P_2} \bar{V}_A^{m,\alpha}(1 - \bar{\alpha}_A^{m,\alpha}T)dP - \bar{H}_A^{m,\beta} - \displaystyle\int_{P_1}^{P_2} \bar{V}_A^{m,\beta}(1 - \bar{\alpha}_A^{m,\beta}T)dP}{\bar{S}_A^{m,\alpha} - \displaystyle\int_{P_1}^{P_2} \bar{V}_A^{m,\alpha}\bar{\alpha}_A^{m,\alpha}dP - \bar{S}_A^{m,\beta} + \displaystyle\int_{P_1}^{P_2} \bar{V}_A^{m,\beta}\bar{\alpha}_A^{m,\beta}dP} \qquad [7\text{-}25a]$$

Assuming $\overline{\alpha}_A^{m,\alpha} T \approx \overline{\alpha}_A^{m,\beta} T \ll 1$, $\overline{V}_A^{m,\alpha}\, \overline{\alpha}_A^{m,\alpha} \approx \overline{V}_A^{m,\beta}\, \overline{\alpha}_A^{m,\beta} \ll (\overline{S}_A^{m,\alpha} - \overline{S}_A^{m,\beta})$, and that $\overline{V}_A^\alpha$ and $\overline{V}_A^\beta$ are independent of pressure,

$$T = \frac{\overline{H}_A^{m,\alpha} + (\overline{V}_A^\alpha - \overline{V}_A^\beta)(P_2 - P_1) - \overline{H}_A^{m,\beta}}{\overline{S}_A^{m,\alpha} - \overline{S}_A^{m,\beta}} \qquad [7\text{-}25b]$$

7.6 GIBBS PHASE RULE

A discussion of phase equilibrium is not complete unless the concept of the *Gibbs phase rule* is introduced. A simple way to visualize the concept is to consider it analogous to the simultaneous solution of a set of mathematical equations. A three variable set, for example, requires three equations if the values of the three variables are to be determined. If only two equations or relationships are known between the three variables, the set can be solved by fixing one of the variables. Simultaneous solution of a two variable-two equation set can then be accomplished. The choice or arbitrary selection of *one* of the variables, in terms of the phase rule concept, means that there is one degree of *freedom* or *variance*, F. The concept is formalized by

$$F = [\text{Number of variables}] - [\text{Number of equations}] \qquad [7\text{-}26]$$

For two equations in a three variable set, $F = 3 - 2 = 1$ degree of freedom.

The concept can now be applied to point X in the $\alpha + \beta$ region of Figure 7.4. At constant temperature, X_A^α is known thus X_B^α is fixed. Likewise, X_A^β is known thus X_B^β is fixed. As a result, there are two composition variables, one for each phase. A generalized expression for the number of composition variables is

$$[\text{Number of composition variables}] = \phi(\Gamma - 1) \qquad [7\text{-}27]$$

where ϕ is the number of phases and Γ is the number of components in the system. Using [7-27], [Number of composition variables] $= 2(2 - 1) = 2$, which is in agreement with the above at point X.

The number of equations that prevail must now be established. From [7-15] and [7-16], the chemical potential of each component is the same in each phase at equilibrium. This leads to two equations. A generalized expression for the number of equations is therefore

$$[\text{Number of equations}] = \Gamma(\phi - 1) \qquad [7\text{-}28]$$

Using [7-28], [Number of equations] $= 2(2 - 1) = 2$. Substituting [7-27] and [7-28] into [7-26],

$$F = \phi(\Gamma - 1) - \Gamma(\phi - 1) = \Gamma - \phi \qquad [7\text{-}29]$$

Since pressure and temperature are additional variables, a general form of the Gibbs phase rule is

$$F = \Gamma - \phi + 2 \qquad [7\text{-}30]$$

Since phase diagrams are normally obtained experimentally at 1 atm pressure, the more common form of [7-30] is

$$F = \Gamma(\text{Components}) - \phi(\text{Phases}) + 1 \qquad [7\text{-}31]$$

For point X in Figure 7.4, $F = 2 - 2 + 1 = 1$ degree of freedom. This means if one variable is fixed (temperature), the other variable (composition of both phases) is determined by the diagram. While the example in this discussion applies to a two component system, [7-31] applies in general to multicomponent systems.

7.7 DISCUSSION QUESTIONS

(7.1) Rewrite [7-11] for the case where $\Delta C_p \neq 0$.

(7.2) What is the basis for determining boundaries on phase diagrams by TL analysis? Briefly discuss and illustrate with a sketch.

(7.3) Two phases in equilibrium contain the same chemical component i. Is the concentration of i in each phase the same? Explain.

(7.4) Two Fe-Mg silicate minerals, garnet and pyroxene, are determined to be in chemical equilibrium. What information is needed to relate the Fe and Mg activity coefficients and concentrations in each mineral?

(7.5) Consider Figure 7.3. Does $\Omega^l = \Omega^\alpha = \Omega^\beta$? Discuss.

(7.6) Based on [7-25b], what condition exists if temperature is independent of pressure?

7.8 EXERCISE PROBLEMS

[7.1] Using the Bi-Pb phase diagram in Appendix E, Figure E.8, calculate the activity of Bi in an equimolar liquid solution of Bi and Pb at 625 K. Assume regular solution behavior.

Ans: $a_{Bi} = 0.38$.

[7.2] Using the results from Exercise Problem [7.1], calculate the equilibrium $P_{O_2(g)}$ over an equimolar Bi-Pb liquid solution at 625 K. The Gibbs free energy of formation of dibismuth trioxide, Bi_2O_3, is $\Delta G^{0,f}_{Bi_2O_3} = -407{,}250$ J/mol at 625 K (Wicks and Block, 1963, p. 21).

Ans: $P_{O_2(g)} \approx 7.4 \times 10^{-23}$ atm.

[7.3] From Kubaschewski and Alcock (1979), solution data for a 70 a/o Bi-Pb solution (Exercise Problem [7.1]) are: $\overline{H}^{m,l}_{Bi} = -70 \times 4.184 = -292.9$ J/mol and $\overline{S}^{XS}_{Bi} = 0.03 \times 4.184 = 0.1255$ J/(mol·K). Using this data, estimate the heat of fusion of Bi.

Ans: $\Delta H^f_{Bi} = 9507$ J/mol or a 13% error—attributed to reading error from the diagram and the assumption that $\Delta C_p \approx 0$.

[7.4] Repeat Exercise Problem [6.9] using the following solubility data estimated from the Pb-Sb phase diagram in Appendix E, Figure E.9: $X_{Sb} = 0.3$ at T = 598 K and $X_{Sb} = 0.2$ at $T = 533$ K. Note that the solubility data is given at a much lower temperature than the refining temperature. Assume:

(1) Calculated solution parameters hold at the higher temperature,

(2) The process involves the reaction

$$2\overline{Sb} + 3PbO(s) \rightarrow 3\overline{Pb} + Sb_2O_3(s);$$

(3) Sb behaves as a dilute solution component at these concentrations.

Ans: $X_{Sb} \approx 0.011$ or ≈ 6500 ppm Sb. Note: assuming Sb behaves ideally (Exercise Problem [6.9]—Method 2) yields a result 23% higher.

[7.5] For dilute solutions of carbon (graphite) in liquid sodium, solubility data was used to determine the partial molar free energy of solution as a function of temperature according to Johnson (1964, p. 23) as follows:

$$\overline{G}_C^l - G_C^{0,S} = 5272 + 68.62T + 8.3144T \ln(X_C).$$

(a) Determine $\overline{H}_C^{XS}$ and $\overline{S}_C^{XS}$.

Ans: $\overline{H}_C^{XS} = h_C = 5272$ J/mol; $\overline{S}_C^{XS} = s_C = -68.62$ J/(mol·K).

(b) The purification of liquid sodium with respect to carbon can be accomplished by *gettering* carbon with calcium at 920°C. Predict the carbon content of the liquid after addition of excess calcium. The solubility of Ca in Na is low, hence, it has no effect on the performance of sodium as a heat transfer medium.

Ans: $X_C = 1.5 \times 10^{-6}$ or ≈ 0.8 ppm C.

[7.6] Using the following data from Kubaschewski and Alcock (1979, p. 386-389) for Cu in a Ag-Cu liquid alloy at 1423 K:

X_{Cu}	0.0	0.1	0.3	0.5	0.7	0.9
$\overline{H}_{Cu}^{m,l}$ (J/mol)	23,014	15,692	7482	3766	1435	159
$\overline{S}_{Cu}^{XS}$ [J/(mol·K)]	5.980	3.084	0.456	0.100	0.084	0.004

(a) Calculate $\overline{G}_{Cu}^{XS}$.

Ans:

X_{Cu}	0.0	0.1	0.3	0.5	0.7	0.9
$\overline{G}_{Cu}^{XS}$ (J/mol)	14,504	11,303	6833	3624	1315	153.3

(b) Calculate $\overline{G}_{Cu}^{XS}$ from the results in Example Problem 7-5 and compare with part (a).

Ans:

X_{Cu}	0.0	0.1	0.3	0.5	0.7	0.9
$\overline{G}_{Cu}^{XS}$ (J/mol)	12,889	10,440	6316	3222	1160	129

These results differ from those of part (a) by 11-16%. See Example Problem 7-5 for further discussion.

[7.7] Using the data from Example Problem 7-8 and Appendix B, Table B.1, predict the eutectic temperature shift resulting from a pressure increase from 1 to 1000 atm.

Ans: From Appendix E, Figure E.7, $T_e = 1053$ K. The eutectic temperature shift is a maximum of + 20 K. In reality, the shift is less—depending upon the actual value of $\overline{V}_{Cu}^\beta$.

[7.8]   Refer to the eutectic phase diagram shown in Appendix E, Figure E.10.
   (a) Identify the components that constitute this system.
        Ans: $NaAlSi_3O_8$ and SiO_2.
   (b) Is the system isobaric or isothermal?
        Ans: Isobaric.
   (c) Give the phase rule expression that is applicable to this diagram.
        Ans: $F = \Gamma - \phi + 1$, pressure is constant.
   (d) Label the diagram at phase assemblages which are: *invariant* ($F =$ 0), *univariant* ($F = 1$), and *divariant* ($F = 2$).
   (e) What is the equation (temperature) of the horizontal line through the eutectic point?
        Ans: $T \approx 1060°C$.
   (f) What is the equation (temperature) of the phase boundary between tridymite and cristobalite?
        Ans: $T \approx 1470°C$.

[7.9]   Using Figure E.11 in Appendix E, develop a temperature-dependent expression for the diamond-graphite phase boundary between 45–105 kbar. Is this boundary invariant, univariant, or divariant? See Exercise Problem [7.8].
        Ans: $P(kbar) = 0.029 \times T(K) + 7$, univariant ($F = 1$).

[7.10] Estimate the equilibrium vapor pressure of $SO_2(g)$ over $SO_2(l)$ at 265 K. State assumptions.
        Ans: $P_{SO_2(g)} = 1.09$ atm, ΔH^{Tr} is assumed constant.

[7.11] Estimate the solid-liquid isothermal transformation temperature of Au at 75 atm. State assumptions.
        Ans: $T_{75\ atm} = 1337$ K, ΔH^{Tr} and ΔV^{Tr} are assumed constant.

ACTIVITY QUOTIENT AND EQUILIBRIUM CONSTANT: APPLICATIONS

8.1 TL DERIVATION OF THE ACTIVITY QUOTIENT AND EQUILIBRIUM CONSTANT

The concept of the equilibrium constant, K_{eq}, leads to the introduction of a new thermodynamic variable. For most thermodynamic problems in simple systems, it is recommended that the thermodynamic loop approach be utilized. The use of the TL insures that the problem is correctly structured. The equilibrium constant concept, however, is most useful in complex multicomponent systems where numerous equilibria must be considered simultaneously. TL analysis is used in Example Problem 8-1 to derive the *activity quotient* and *equilibrium constant* of a chemical reaction. Several applications are presented throughout this chapter.

Example Problem 8-1

Using TL analysis and the general chemical reaction

$$n_A A + n_B B \rightarrow n_C C + n_D D,$$

derive:

(a) Activity quotient

$$J_a = \frac{a_C^{n_C} \cdot a_D^{n_D} \ (\text{Products})}{a_A^{n_A} \cdot a_B^{n_B} \ (\text{Reactants})};$$

(b) Equilibrium constant

$$K_{eq} = \frac{a_C^{n_C} \cdot a_D^{n_D} \ (\text{Products})}{a_A^{n_A} \cdot a_B^{n_B} \ (\text{Reactants})};$$

What is the relationship between J_a and K_{eq}?

Solution

(a) The activity of each component is shown in parentheses.

(1) Set Up.

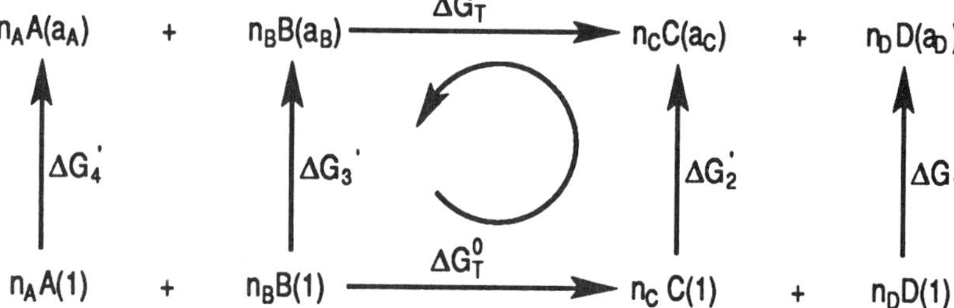

(2) Sum.

$$\Sigma \Delta G_{TL} = 0 = \Delta G_T^0 + \Delta G_1' + \Delta G_2' - \Delta G_T - \Delta G_3' - \Delta G_4'.$$

(3) Substitute. Applying [6-3],

$$\Delta G_1' = n_D RT \ln(a_D); \quad \Delta G_2' = n_C RT \ln(a_C).$$
$$\Delta G_3' = n_B RT \ln(a_B); \quad \Delta G_4' = n_A RT \ln(a_A).$$

Substituting into $\Sigma \Delta G_{TL} = 0$,

$$0 = \Delta G_T^0 + n_D RT \ln(a_D) + n_C RT \ln(a_C) - \Delta G_T - n_B RT \ln(a_B) - n_A RT \ln(a_A).$$

(4) Solve. Collecting like terms and rearranging,

$$\Delta G_T = \Delta G_T^0 + RT \ln \left[\frac{a_C^{n_C} \cdot a_D^{n_D} \text{ (Products)}}{a_A^{n_A} \cdot a_B^{n_B} \text{ (Reactants)}} \right].$$

The activity quotient, J_a, is defined by

$$J_a = \frac{a_C^{n_C} \cdot a_D^{n_D} \text{ (Products)}}{a_A^{n_A} \cdot a_B^{n_B} \text{ (Reactants)}}.$$

Hence,

$$\Delta G_T = \Delta G_T^0 + RT \ln(J_a) \qquad [8\text{-}1]$$

For a reaction involving n_i moles of component i at activity a_i, the activity quotient becomes

$$J_a = \frac{\prod a_i^{n_i} \text{ (Products)}}{\prod a_i^{n_i} \text{ (Reactants)}} \qquad [8\text{-}2]$$

where $\prod$ is defined as the *product of activities*.

(b) At equilibrium, $\Delta G_T = 0$ and $J_a = K_{eq}$. Applying [8-1],

$$\Delta G_T^0 = -RT \ln(K_{eq}) \qquad [8\text{-}3]$$

The reader should note the similarity between [8-3] and [6-3]. For example, applying [6-14] to the reaction

$$A^0(s) \rightarrow \overline{A}(s),$$

$$\overline{G}_A^{m,s} = RT \ln(a_A) \qquad [8\text{-}4]$$

However, applying [8-3] for the same reaction at *equilibrium*,

$$\Delta G_T^0 = -RT \ln(a_A) \qquad [8\text{-}5]$$

8.2 EFFECT OF TEMPERATURE AND PRESSURE AT EQUILIBRIUM

Two cases define the relationship between the equilibrium constant, temperature, and pressure.

Case I. K_{eq} and Temperature:

Substituting [8-3] into [4-30],

$$\frac{\partial}{\partial T}\left[\frac{-RT\ln\left(K_{eq}\right)}{T}\right]_P = -\frac{\Delta H_T^0}{T^2}$$

or

$$\frac{d\ln(K_{eq})}{dT}\bigg|_P = \frac{\Delta H_T^0}{RT^2} \qquad [8\text{-}6]$$

[8-6] is known as the *Van't Hoff equation*. Assuming ΔH_T^0 is constant ($\Delta C_p = 0$) and integrating,

$$\ln(K_{eq}) = -\frac{\Delta H_T^0}{RT} + C \qquad [8\text{-}7]$$

where C is an integration constant. From inspection of [8-7], if $\Delta H_T^0 > 0$ (endothermic reaction), K_{eq} increases as temperature increases and the reaction tends to shift in the direction of the products until a new K_{eq} is established. Conversely, if $\Delta H_T^0 < 0$ (exothermic reaction), K_{eq} decreases as temperature increases and the reaction tends to shift toward the reactants until a new K_{eq} is established. Note from [8-7] that the standard enthalpy of a reaction can be obtained directly from experimental measurement of K_{eq} as a function of temperature.

Case II. K_{eq} and Pressure:

Based on [8-3], K_{eq} is a function of temperature only since ΔG_T^0 is defined at constant pressure ($P_T = 1$ atm). According to LeChatelier's principle, if a reaction at equilibrium is subjected to a change in pressure, the reaction will shift in a direction so as to relieve the pressure change. A simple example is the shift in mole fraction of a gas that occurs as a result of a pressure change for the reaction

$$A(s) + 2B(g) \rightarrow AB_2(g).$$

Substituting $a_i = Y_i P_T$ into [8-2] and assuming that the pressure effect on volume of solid A is negligible in comparison to the gas species,

$$K_{eq} = \frac{a_{AB_2(g)}}{a_{B(g)}^2} = \frac{Y_{AB_2(g)} \cdot P_T}{(Y_{B(g)} \cdot P_T)^2} = \frac{Y_{AB_2(g)}}{Y_{B(g)}^2 P_T}.$$

If P_T increases, the above equality shows that the ratio $Y_{AB_2(g)} / Y_{B(g)}^2$ increases in order to *maintain* K_{eq} *constant*. This result is consistent with LeChatelier's principle since the reaction shifts from left to right and $Y_{AB_2(g)}$ increases.

8.3 TL ANALYSIS AND K_{eq}: EXAMPLE PROBLEMS

In this section, example problems utilizing the equilibrium constant and TL analysis are presented. Some K_{eq} examples are solved using K_{eq} directly without applying the TL.

Example Problem 8-2

Using the same data presented in Example Problem 5-2, calculate P_{H_2} / P_{H_2O} from [8-3].

Solution

Inserting the data from Example Problem 5-2 into [8-3],

$$\Delta G_{1273}^0 = -173,039 = -8.3144(1273)\ln(K_{eq})$$

$$= -8.3144(1273) \times \ln(P_{H_2}^2 / P_{H_2O}^2)$$

where

$$a_{Cr_2O_3(s)} = a_{Cr(s)} = 1.$$

Hence,

$$P_{H_2} / P_{H_2O} = \sqrt{K_{eq}} = \underline{3.55 \times 10^3}.$$

Example Problem 8-3

The carburized steel in Example Problem 7-6 is exposed to an atmosphere of CO_2 and CO. It is suggested that decarburization may occur at 500°C. Under what circumstance is this a valid concern? Assume $P_T = 1$ atm.

Solution

At 500°C, decarburization involves dissociation of Fe_3C according to the reaction

$$Fe_3C(s) + CO_2(g) \rightarrow 3Fe(s) + 2CO(g).$$

The amount of carbon dissolved in ferrite is assumed to be insignificant compared to the carbon in carbide form. ΔG_T^0 (J/mol) is obtained from the data in Table A.4:

(1) $Fe_3C(s) \rightarrow 3Fe(s) + C(s)$ $\qquad\qquad\qquad \Delta G_T^0 = -26,700 + 24.77T$

(2) $CO_2(g) \rightarrow C(s) + O_2(g)$ $\qquad\qquad$ $\Delta G_T^0 = 394{,}170 + 0.84T$

(3) $2C(s) + O_2(g) \rightarrow 2CO(g)$ $\qquad\qquad$ $\Delta G_T^0 = -223{,}440 - 175.32T$

Combining (1), (2), and (3) results in the decarburization reaction for which

$$\Delta G_T^0 \text{ (J/mol)} = 144{,}030 - 149.71T.$$

At 773 K, $\Delta G_{773}^0 = 28{,}304$ J/mol. Inserting this data into [8-3],

$$28{,}304 = -8.3144(773)\ln(K_{eq}) = -8.3144(773)\ln(P_{CO}^2 / P_{CO_2})$$

where $a_{Fe} = a_{Fe_3C} = 1$. Hence,

$$P_{CO}^2 / P_{CO_2} = 1.22 \times 10^{-2}.$$

If $P_{CO} + P_{CO_2} = 1$, simultaneous solution gives $P_{CO} = 0.10$ atm, and $P_{CO_2} = 0.90$ atm. Decarburization at 500°C is a valid concern if $P_{CO}^2 / P_{CO_2} < 0.0122$.

Example Problem 8-4

Referring to Example Problem 7-4, estimate the carbon concentration that remains dissolved as graphite at 660°C just prior to forming the carbide Al_4C_3.

Solution

TL analysis is recommended for this problem because partial molar Gibbs free energy is obtained directly from the reference problem. It is unnecessary to convert to activity. For practical purposes, Al is a pure component.

(1) Set Up.

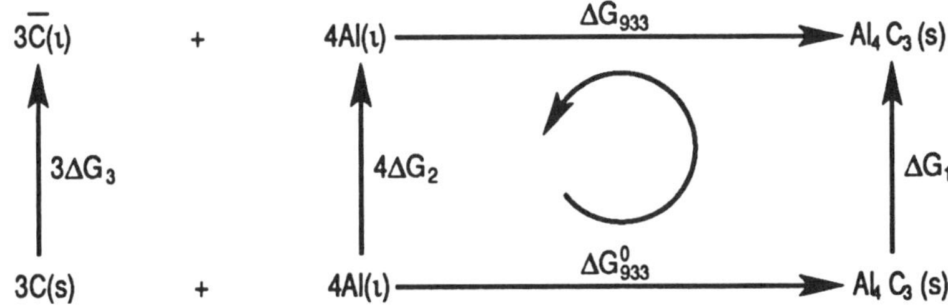

The maximum carbon concentration is established when $\Delta G_{933} = 0$.

(2) Sum.

$$\Sigma \Delta G_{TL} = 0 = \Delta G_{933}^0 + \Delta G_1 - \Delta G_{933} - 4\Delta G_2 - 3\Delta G_3.$$

(3) Substitute.

$\Delta G_{933}^0 = -266{,}521 + 96.23(933) = -176{,}738$ J/mol from Table A.4.

$\Delta G_1 = 0$; $4\Delta G_2 \approx 0$ ($\overline{Al}$(l) is almost pure).

$\Delta G_{933} = 0$ (equilibrium).

$3\Delta G_3 = 3(\overline{G}_C^l - G_C^{0,s}) = 3\{h - T[s - R\ln(X_C^l)]\}$

$= 3\{239{,}490 - 933[101 - 8.3144\ln(X_C^l)]\}.$

Substituting the above data into $\Sigma\Delta G_{TL} = 0$,

$0 = -176{,}738 - 3\{239{,}490 - 933[101 - 8.3144\ln(X_C^l)]\}.$

(4) Solve.

From [6-5b], $X_C^l = 3.7 \times 10^{-12} \approx \underline{0.002\ ppb}$.

From a practical point of view, the formation of $Al_4C_3(s)$ is certain.

Example Problem 8-5

In this example problem, thermodynamic analysis of a three-component closed system is presented to illustrate the utility of the equilibrium constant. For example, in Na-C-O equilibria, seven chemical reactions are analyzed to characterize seven condensed phases and three gases. Selecting the following reaction for illustration:

$$Na_2CO_3(s) + CO(g) = 2Na(g) + 2CO_2(g),$$

(a) Calculate P_{CO} as a function of P_{Na}.

Solution

At equilibrium,

$$K_{eq} = P_{CO_2}^2 \cdot P_{Na}^2 / P_{CO}.$$

Assuming $P_T = P_{CO} + P_{CO_2} + P_{Na}$, and substituting P_{CO_2} into the above expression for K_{eq},

$$P_{CO} = \frac{K_{eq} + 2P_T P_{Na}^2 - 2P_{Na}^3 \pm \sqrt{K_{eq}(K_{eq} - 4P_{Na}^3 + 4P_T P_{Na}^2)}}{2P_{Na}^2}.$$

P_{CO} is calculated from selected values of P_T and P_{Na}. K_{eq} is obtained from [8-3] at constant temperature.

(b) Plot the result from part (a) at $P_T = 1$ atm and $T = 1000$ K.

Solution

At $P_T = 1$ atm and $T = 1000$ K, $\Delta G_T^0 = 258{,}847$ J/mol.

Hence,

$$K_{eq} = \exp(-\Delta G_T^0 / RT) = \exp(-258{,}847/8.3144 \times 1000)$$
$$= 3.0157 \times 10^{-14}.$$

Substituting K_{eq} and $P_T = 1$ atm into the equation for P_{CO} and plotting P_{CO} versus P_{Na} results in the Na_2CO_3 *stability diagram* shown in Figure 8.1. Dissociation of Na_2CO_3 occurs to the left of the Na_2CO_3 equilibrium line.

Correlation of several equilibria derived in a similar fashion allows development of ternary isotherms. A detailed discussion of this system is given by Weaver et al. (1977) and by Gokhale and Johnson (1982).

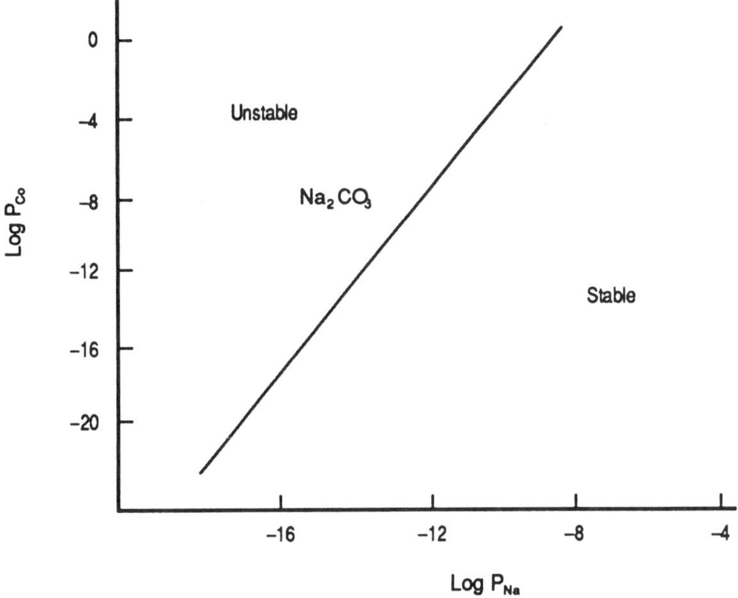

Figure 8.1 Calculated stability diagram for Na_2CO_3 according to the reaction $Na_2CO_3(s) + CO(g) = 2Na(g) + 2CO_2(g)$. $P_T = 1000$ atm; $T = 1000$ K.

Example Problem 8-6

Referring to Example Problem 6-9: (a) Repeat the problem using the definition of the equilibrium constant; (b) Express the concentration of hydrogen in solution, C_H, as a function of the heat of solution.

Solution
(a) The dissociation reaction is

$$1/2H_2(g) \rightarrow \overline{H} \, ;$$

$$K_{eq} = a_H / P_{H_2}^{1/2} = (C_H / C_H^0) / P_{H_2}^{1/2} \Rightarrow C_H = C_H^0 K_{eq} P_{H_2}^{1/2} \, .$$

Hence,

$$C_H = k P_{H_2}^{1/2} \qquad\qquad [6\text{-}72b]$$

where $k = C_H^0 K_{eq}$.

(b) From [8-7],

$$\ln(K_{eq}) = -\frac{\Delta H_s^0}{RT} + C \qquad [8\text{-}8]$$

where $\Delta H_s^0 = \overline{H}_{\mathrm{H}} - (1/2)H_{\mathrm{H}_2}^{0,g}$, the *molar heat of solution* relative to gaseous hydrogen. Solving [8-8] for K_{eq},

$$K_{eq} = (C_{\mathrm{H}}/C_{\mathrm{H}}^0)/P_{\mathrm{H}_2}^{1/2} = e^{[(-\Delta H_s^0/RT)+C]} = e^C \cdot e^{-\Delta H_s^0/RT}.$$

Hence,

$$C_{\mathrm{H}} = C_H^0 e^{-\Delta H_s^0/RT} \qquad [8\text{-}9]$$

where $C_0 = C_{\mathrm{H}}^0 e^C P_{\mathrm{H}_2}^{1/2}$.

8.4 THE NERNST EQUATION

A definition of Electromotive Force E (EMF) for oxidation or charge transfer is given by $\Delta G = nFE$ where ΔG is the Gibbs free energy change, n is valence, and F is *Faraday's constant* (96,500 coulomb/eq). The *Nernst equation* is obtained by substituting this expression into [8-1] and dividing through by nF. The result is

$$E_{\mathrm{Cell}} = E_{\mathrm{Cell}}^0 + 2.303(RT/nF)\log_{10}(J_a) \qquad [8\text{-}10]$$

For corrosion reactions which normally occur at room temperature (298 K), [8-10] is simplified as follows:

$$2.303(RT/nF) = \frac{2.303 \times [8.3144\ \mathrm{J/(mol \cdot K)}] \times 298\,\mathrm{K}}{(n\ \mathrm{eq/mol}) \times 96,500\ \mathrm{coulomb/eq}} = \frac{0.0592}{n}\,\mathrm{J/coulomb;}$$

substituting into [8-10],

$$E_{\mathrm{Cell}} = E_{\mathrm{Cell}}^0 + (0.0592/n)\log_{10}(J_a) \qquad [8\text{-}11]$$

The standard state is chosen such that for dilute concentrations, in corrosion, $a_i = C_i = 1$ mole/liter (1 *molar solution*).

Example Problem 8-7

An *electrochemical cell* consists of Sn and Fe electrodes in an electrolytic solution at standard concentrations of Sn and Fe ions.

(a) Which electrode will act as the anode and hence be the corroding one? Start by assuming that Fe is anodic to Sn at $T = 25°C$.

Solution

The cell reaction is the sum of anodic (oxidation) and cathodic (reduction) reactions:

(1) Fe $\to$ Fe^{+2} + 2e$^-$ (Anode)
(2) Sn^{+2} + 2e$^-$ $\to$ Sn   (Cathode)
(1) + (2) = (3): Fe + Sn^{+2} $\to$ Fe^{+2} + Sn   (Cell)

From [8-2],

$$J_a = \frac{C_{Fe^{+2}} \cdot C_{Sn}}{C_{Fe} \cdot C_{Sn^{+2}}} = \frac{1 \cdot 1}{1 \cdot 1} = 1$$

Metallic Sn and Fe are present as pure solids, hence $C_{Sn} = C_{Fe} = 1$ and $C_{Sn^{+2}} = C_{Fe^{+2}} = 1$ mol/liter. Equation [8-11] becomes $E_{Cell} = E^0_{Cell}$ since $\log_{10}(J_a) = 0$. The standard EMF for each *half cell* (anode and cathode) is obtained from Appendix C, Table C.1, as follows:

(1) $Fe \rightarrow Fe^{+2} + 2e^-$ $\qquad\qquad\qquad$ $E^0 = -0.44$ volts $\qquad$ (Half Cell)
(2) $Sn^{+2} + 2e^- \rightarrow Sn$ $\qquad\qquad\qquad$ $E^0 = + 0.14$ volts $\qquad$ (Half Cell)

(1) + (2) = (3) $Fe + Sn^{+2} \rightarrow Fe^{+2} + Sn$ $\quad$ $E^0_{Cell} = -0.30$ volts. (Cell)

Since the cell voltage is negative, $\Delta G_{Cell} = nFE_{Cell} < 0 \Rightarrow$ the cell reaction will occur as written, i.e., <u>Fe is anodic to Sn</u>.

(b) If the ratio $C_{Fe^{+2}}/C_{Sn^{+2}} = 10^{15}$, is Fe anodic to Sn?

Solution
Applying [8-11] where $E^0_{Cell} = -0.30$ volts from part (a),

$$E_{Cell} = -0.30 + (0.0592/2)\log[(C_{Fe^{+2}} \cdot C_{Sn})/(C_{Sn^{+2}} \cdot C_{Fe})]$$
$$= -0.30 + (0.0592/2)\log(10^{15}) = 0.144 \text{ volts.}$$

Since E_{Cell} is now positive, the polarity is reversed ($\Delta G_{Cell} = nFE_{Cell} > 0$), and <u>Sn is anodic to Fe</u>. This illustrates the fact that a change in composition of the electrolyte can localize corrosion in unexpected areas. An evaluation based only on the EMF series often leads to error.

8.5 DISCUSSION QUESTIONS

(8.1)   Explain the difference between activity quotient and equilibrium constant. Under what condition are they equal?

(8.2)   Applying $\Delta G_T = \Delta G^0_T + RT \ln(J_a)$, are the reactants and products in equilibrium? Is this equation applicable to pure phases only? Why?

(8.3)   What isobaric relationship relates the equilibrium constant of a reaction to the enthalpy change and temperature of the reaction?

(8.4)   Explain how the equilibrium constant can be used to determine whether a reaction will tend to shift toward products or reactants.

(8.5)   Explain how the EMF of a cell can be determined from half cell reactions. What law is this an example of?

(8.6)   Two electrodes, A and B, are immersed in an electrolyte. If it is assumed that A is anodic to B and it is then determined that $E_{Cell} > 0$, will A corrode?

(8.7)   Why is TL analysis applicable to problems utilizing the equilibrium constant?

(8.8)   Develop [8-9] directly from [6-72a]. What does a comparison of [8-9] and [6-72a] suggest concerning the integration constant in [8-8]?

8.6 EXERCISE PROBLEMS

[8.1] Referring to Example Problem 5-6: (a) Find the equilibrium constant for the reaction and discuss the effect of increasing (b) total pressure and (c) temperature.

> *Ans:* (a) $K_{eq} = Y_{HCl}^4 P_T /(Y_{H_2}^2 Y_{SiCl_4})$, (b) The yield of Si(s) and HCl(g) decreases because $Y_{HCl(g)}$ must decrease in order to maintain K_{eq} constant; (c) K_{eq} increases because the reaction is endothermic.

[8.2] The decomposition of $Fe_3C(s)$ (*decarburization*) in a steel exposed to a hydrogen environment can result in the internal formation of $CH_4(g)$. Internal pressure created by accumulation of the CH_4 can lead to so called hydrogen attack because CH_4 cannot diffuse from the lattice structure. Repeat Example Problem 8-3 for exposure to a mixture of $H_2(g)$ and $CH_4(g)$ at 500°C and determine the minimum $P_{CH_4}/P_{H_2}^2$ ratio that can exist without causing significant decarburization and subsequent hydrogen attack.

> Ans: $P_{CH_4}/P_{H_2}^2 = 7.283$.

[8.3] Estimate the total enthalpy change associated with modified austempering of one gram mole of 1050 steel. Identify all assumptions made and source materials used.

> *Ans:* $\Delta H_{Total} = 5091$ J/mol.

[8.4] Manganese sulfide inclusions in steel are interfacial traps (sinks) for hydrogen and hence tend to increase the total or "effective" hydrogen solubility. Depending upon the distribution and size of the inclusions, they may increase the susceptibility of the steel to *hydrogen embrittlement* or *hydrogen environment cracking* (HEC).

(a) If $H_2S(g)$ and $H_2(g)$ are present in the contacting atmosphere, what is the critical ratio P_{H_2S}/P_{H_2} above which MnS(s) and $H_2(g)$ form at 1500°C and 1 atm?

> *Ans:* $P_{H_2S}/P_{H_2} = 5 \times 10^{-5}$.

(b) The calculated critical ratio in (a) for minimizing HEC in steel may not be confirmed experimentally. Name two factors that are not considered in these calculations.

> *Ans:* Solid state kinetics are normally slow and are not involved in thermodynamic calculations. The calculated ratio does not take into account solubility effects.

[8.5] Referring to the data in Appendix A, Table A.5, show that

$$\ln(P_{Cu}^l)(atm) = -40,349/T - 1.21\ln(T) + 23.79.$$

[8.6] Referring to the paper by Iwai, Takahashi, and Handa (1986):
(a) Confirm the data in the last three columns of Table II.
(b) Equilibrium data are found in Tables II and III. Use the least mean

squares treatment to derive the expression:

$$\Delta G^{0,f}_{Mo_2C} \text{ (J/mol)} = -68{,}270 + 8.23T$$

where $1173 \text{ K} \le T \le 1573 \text{ K}$.

(c) What assumption is implied but not directly mentioned in the article?

    Ans: $\Delta C_p \approx 0$.

[8.7]   The vapor pressure data in Appendix A, Table A.5, can be expressed in the form

$$\log_{10}(P) = AT^{-1} + B\log_{10}(T) + D$$

where $C = 0$.

(a) Starting with [7-20], develop the above expression for the pressure of a vapor over its coexisting condensed phase at equilibrium.

(b) What is the form of the equation if $\Delta C_p = 0$?

    Ans: $\log_{10}(P) = AT^{-1} + D$.

[8.8]   Confirm that hydrogen dissolved in Ti(α) follows Sievert's law. Use the following data given at 610°C by McQuillan (1950):

$P_{H_2(g)}$ (atm)	Hydrogen Solubility in Ti(α) (ppm)
5.263×10^{-5}	235
5.263×10^{-4}	750
2.632×10^{-3}	1705

    Ans: k values are constant within experimental error, thereby confirming the law.

[8.9]   Repeating Example Problem 7-6 for carburization of a low carbon steel in a mixture of CO and CO_2 at 900°C, calculate the carburizing potential required to develop 0.8 w/o C at the surface.

    Ans: $P^2_{CO} / P_{CO_2} = 19.466$.

[8.10]   The solubility of hydrogen in liquid binary Al-Li alloys is given by Anyalebechi, Talbot, and Granger (1988):

Pure Al: $\log(s/s^0) - (1/2)\log(P/P^0) = -2700/T + 2.720$

Al-1 Pct Li: $\log(s/s^0) - (1/2)\log(P/P^0) = -2113/T + 2.568$

Al-2 Pct Li: $\log(s/s^0) - (1/2)\log(P/P^0) = -2797/T + 3.329$

Al-3 Pct Li: $\log(s/s^0) - (1/2)\log(P/P^0) = -2889/T + 3.508$

where $s^0 = 1 \text{ cm}^3/100$ gm metal at STP and $P^0 = 1.01325 \times 10^5$ Pa. The equations are valid from 913 to 1073 K and from 5.3×10^4 to 10.7×10^4 Pa. Determine the enthalpy of solution at each concentration. The last term in each expression includes an activity correction; hence, it cannot be used to calculate ΔS°_s directly.

Ans:

Pct Li	ΔH_s^0 (J/mol H_2)
0	103,400
1	80,920
2	107,114
3	110,638

[8.11] Referring to Exercise Problem [8.10], the dissociation of molecular hydrogen in liquid aluminum alloys is a two step process involving: (1) Dissociation of diatomic hydrogen into monatomic hydrogen and (2) Solution of monatomic hydrogen in the metal. Using enthalpy data for the Al-1 pct Li alloy, calculate ΔH for step 2 if for step 1, $\Delta H_{Ave} = 419,500$ J/mol H_2. ΔH_{Ave}, defined as the enthalpy of dissociation of diatomic hydrogen, is effectively constant over the temperature range of interest. ΔH is the solute binding enthalpy for solution of monatomic hydrogen in the metal.

   Ans: $\Delta H = -169,290$ J/gm atom H.

[8.12] Determine K_{eq} as a function of temperature for the reaction in Exercise Problem [4.22].

   Ans: $K_{eq} = \exp(-455.96 \times 10^{-9}T^2 + 8.46 \times 10^{-3}T - 2102.14 \times T^{-0.5} - 6399.74T^{-1} - 295.92 \times 10^3 T^{-2} - 28.65\ln(T) + 240.59)$ where $T = T(K)$.

[8.13] Referring to Exercise Problem [5.17],

   (a) What is the equilibrium constant for the reaction?

   Ans: $K_{eq} = \left(P_{CO_2} \cdot P_{CH_4}\right)/(a_C^2 \cdot P_{H_2O}^2) = 0.21.$

   (b) In which direction will the reaction shift if Ja increases? If Ja decreases?

   Ans: J_a increases $\Rightarrow$ shift to the left, J_a decreases $\Rightarrow$ shift to the right. The shift occurs until equilibrium is re-established ($J_a = K_{eq}$).

   (c) What would cause the equilibrium constant to change?

   Ans: A change in temperature.

APPENDIX A
TABLES OF THERMODYNAMIC DATA

Table A.1

Standard Heats of Formation, Standard Entropies, Melting (M.P.) and Boiling (B.P.) Points.

Solid < >, Liquid [ ], Gas ( ).

P^0 = 1 atm (1 bar for References 1 and 5 below). Standard T = 298 K.

Substance	$\Delta H^{0,f}_{298}$ (kJ/mol)	S^0_{298} [J/(mol·K)]	M.P. (K)	B.P. (K)	Ref.*
<Al>	0	28.33	933	2773	4
<Al$_2$O$_3$>$_\alpha$	−1675.73	50.90	—	—	1
<Al$_2$SiO$_5$>[a]	−2591.27	91.40	—	—	1
<B>	0	5.86	2323	—	4
<B$_4$C>	−57.70	27.07	2623	—	4
<C>$_{Graphite}$	0	5.69	—	—	4
<Ca>$_\alpha$	0	41.63	1123	1623	4
<CaAl$_2$Si$_2$O$_8$>	−4232.74	199.30	—	—	1
<CaCO$_3$>[b]	−1207.43	87.99	—	—	5
<CaCO$_3$>[c]	−1206.70	88.70	—	—	5
Tremolite[d]	−12302.50	0.55	—	—	1
<CaO>	−633.88	39.75	2873	—	4
<CaSiO$_3$>	−1635.22	82.01	—	—	5
(CH$_4$)	−74.81	186.26	—	—	1
(CO$_2$)	−393.51	213.80	—	—	3
<Cr>	0	23.77	2123	2773	4
<Cr$_2$O$_3$>	−1129.00	81.18	2673	—	3
<Cu>	0	33.35	1356	2843	4
<Cu>	0	33.15	1356	—	5
<Cu$_2$O>	−167.38	93.94	1503	—	4
<Cu$_2$S>$_\alpha$	−80.115	120.75	—	—	2
<Fe>	0	27.28	1809	—	3
<FeO(OH)>	−545.59	75.73	—	—	6
<Fe$_2$O$_3$>	−823.41	89.96	—	—	7
<Fe$_2$P>	−160.25	72.38	—	—	4
<FeS$_2$>	−177.40	53.14	—	—	4
<Fe$_2$SiO$_4$>	−1479.36	148.32	1490	1800	5
<FeTiO$_3$>	−1233.26	108.50	—	—	1
(H$_2$)	0	130.68	—	—	5
[H$_2$O]	−285.90	70.09	273	373	4
(H$_2$O)	−241.81	188.80	—	—	3
(H$_2$S)	−20.627	205.80	—	—	5
(HCl)	−92.32	186.80	159	265	3
[Hg]	0	75.91	234	630	3
<HgO>	90.80	70.30	—	—	4

Table A.1 Continued
Standard Heats of Formation, Standard Entropies,
Melting (M.P.) and Boiling (B.P.) Points.
Solid < >, Liquid [ ], Gas ( ).
P^0 = 1 atm (1 bar for References 1 and 5 below). Standard T = 298 K.

$\langle Mg \rangle$	0	32.51	923	1378	4
$\langle MgO \rangle$	−601.30	27.41	3043	—	4
$\langle Mg(OH)_2 \rangle$	−925.50	63.00	—	—	1
$\langle Mg_3Si_2O_5(OH)_4 \rangle$	−4361.66	221.30	—	—	5
$\langle Mg_3Si_4O_{10}(OH)_2 \rangle$	−5915.90	260.83	—	—	5
(N_2)	0	191.52	—	—	3
$\langle NaAlSi_2O_6 \rangle$	−3029.40	133.47	—	—	5
$\langle NaAlSi_3O_8 \rangle$	−3935.12	207.40	—	—	5
Glaucophane[e]	−11963.86	535.00	—	—	1
(NH_3)	−46.02	192.34	—	—	4
$\langle Ni \rangle$	0	29.79	1726	3183	4
$\langle NiO \rangle$	−240.60	38.08	2233	—	4
(O_2)	0	205.02	—	—	3
(P_2)	143.85	218.00	—	—	3
(S_2)	128.49	228.17	—	—	5
$\langle Si \rangle$	0	18.83	1693	2873	4
$(SiCl_4)$	−662.81	330.86	203	331	3
$[Si_3N_4]$	−749.00	96.24	—	2173	4
$\langle SiO_2 \rangle_\alpha$	−910.70	41.46	—	—	5
(SO_2)	−296.81	248.22	—	—	5
$\langle Ti \rangle_\alpha$	0	30.54	—	—	3
$\langle TiC \rangle$	−183.70	24.27	3423	—	3
$\langle W \rangle$	0	33.48	3653	—	4
$\langle WO_2 \rangle$	−560.70	66.95	—	—	4
$\langle WO_3 \rangle$	−836.90	83.27	1746	—	4
$\langle W_3O_8 \rangle$	−2232.20	—	—	—	4

*References:

   (1) Holland, 1990, Table 7, p. 103–104.
   (2) Kelley, 1960.
   (3) Kubaschewski and Alcock, 1979, Table A, p. 267.
   (4) Kubaschewski and Evans, 1958, Table A, p. 226.
   (5) Robie et al., 1979.
   (6) Schmalz, 1959.
   (7) Wicks and Block, 1963.

Footnotes:

   (a) Andalusite polymorph
   (b) Aragonite polymorph
   (c) Calcite polymorph
   (d) Tremolite: $\langle Ca_2Mg_5Si_8O_{22}(OH)_2 \rangle$
   (e) Glaucophane: $\langle Na_2Mg_3Al_2(Si_4O_{11})_2(OH)_2 \rangle$

Table A.2
Heats of Transformation and Fusion at Specified Transformation (Tr. Pt.) and Melting (M.P.) Points.

P^0 = 1 atm (1 bar for Reference 5 below).

s = solid, l = liquid, g = gas; α, β, δ, and γ are polymorphic solids.

Substance	ΔH^{Tr} (kJ/mol)	Tr. Pt. (K)	M.P. (K)	Ref.[*]
Al	10.461 ($s \rightarrow l$)	—	933	3
Au	12.760 ($s \rightarrow l$)	—	1336	3
Bi	10.879 ($s \rightarrow l$)	—	544	3
Cr	19.25 ($s \rightarrow l$)	—	2123	4
Cu	12.972 ($s \rightarrow l$)	—	1356	3
Cu	13.054 ($s \rightarrow l$)	—	1356	5
Cu_2S	3.849 ($\alpha \rightarrow \beta$)	376	—	2
Cu_2S	0.8368 ($\beta \rightarrow \gamma$)	623	—	2
Cu_2S_γ	10.878 ($s \rightarrow l$)	—	1403	2
Fe	0.669 ($\alpha \rightarrow \gamma$)	1187	—	3
Fe	13.770 ($s \rightarrow l$)	—	1809	4
Fe_2SiO_4	92.173 ($s \rightarrow l$)	—	1490	5
H_2O	41.09 ($l \rightarrow g$)	373	—	4
Li	2.929 ($s \rightarrow l$)	—	453	3
Mn	2.009 ($\alpha \rightarrow \beta$)	993	—	4
Mn	2.301 ($\beta \rightarrow \gamma$)	1373	—	4
Mn	1.800 ($\gamma \rightarrow \delta$)	1409	—	4
Mn_δ	13.390 ($s \rightarrow l$)	—	1517	4
Pb	4.812 ($s \rightarrow l$)	—	600	3
S	0.368 ($\alpha \rightarrow \beta$)	369	—	1

Table A.2 Continued
Heats of Transformation and Fusion at Specified Transformation (Tr. Pt.)
and Melting (M.P.) Points.
P^0 = 1 atm (1 bar for Reference 5 below).
s = solid, l = liquid, g = gas; α, β, δ, and γ are polymorphic solids.

S	1.226 ($\beta \rightarrow l$)	392	—	1
S	20.92 ($l \rightarrow g$)	718	—	1
Sb	19.876 ($s \rightarrow l$)	—	903	3
Si	50.630 ($s \rightarrow l$)	—	1693	4
SO_2	24.937 ($l \rightarrow g$)	263	—	5
Ta	24.558 ($s \rightarrow l$)	—	3253	4
Ti	3.473 ($\alpha \rightarrow \beta$)	1155	—	4
Ti_β	18.830 ($s \rightarrow l$)	—	1933	4

*References

(1) Handbook of Chemistry and Physics, 1989, p. D-45.

(2) Kelley, 1960.

(3) Kubaschewski and Alcock, 1979, Table B, p. 326.

(4) Kubaschewski and Evans, 1958, Table B, p. 286.

(5) Robie et al., 1979.

Table A.3A

Heat Capacities: $C_p = a + bT + cT^{-2} + dT^{-0.5}$. Units: J/(mol·K), T is in degrees K,
M.P. denotes melting point. $P^0 = 1$ atm (1 bar for References 2 and 6 below).
Solid < >, Liquid [ ], Gas ( ).

Substance	a	$b \times 10^3$	$c \times 10^{-5}$	$d \times 10^{-3}$	Temp. Range (K)	Ref.*
<Al>	20.67	12.39	—	—	298–M.P.	4
[Al]	29.29	—	—	—	M.P.–1273	5
<Al$_4$C$_3$>	100.76	132.23	—	—	298–600	5
<Au>	23.68	5.19	—	—	298–M.P.	4
[Au]	29.29	—	—	—	M.P.–1600	4
<BN>	7.62	15.15	—	—	298–1200	5
<C>[a]	17.15	4.27	–8.79	—	298–2300	5
<Ca>$_\alpha$	21.92	14.64	—	—	298–723	1
<Trem>[b]	1214.40	26.53	–123.62	–7.39	298–	2
(CH$_4$)	23.64	47.87	–1.92	—	298–1500	4
<Cr>	24.44	9.88	–3.68	—	298–M.P.	4
[Cr]	39.33	—	—	—	M.P.–	4
<Cr$_2$O$_3$>	119.40	9.21	–15.65	—	350–1800	4
<Cu$_2$S>$_\alpha$	81.588	—	—	—	298–376	3
<Cu$_2$S>$_\beta$	97.278	—	—	—	376–623	3
<Cu$_2$S>$_\gamma$	85.019	—	—	—	623–M.P.	3
<Fe>$_\alpha$	17.49	24.77	—	—	273–1181	5
<Fe>$_\gamma$	7.70	19.50	—	—	1181–1674	5
<FeS$_2$>	74.81	5.52	–12.76	—	298–1000	4
<FeTiO$_3$>	116.61	18.24	–20.04	—	298–1640	4
[H$_2$O]	75.44	—	—	—	273–373	4
(H$_2$O)	30.00	10.71	0.33	—	298–2500	4
(H$_2$S)	29.37	15.40	—	—	298–1800	5
<Jd>[c]	301.13	10.143	–22.393	–2.055	298–1300	6
<MgO>	42.59	7.28	–6.19	—	298–2100	5
<Mg(OH)$_2$>	158.40	–4.076	–10.523	–1.1713	298–	2
<Mn>$_\alpha$	21.59	15.94	—	—	298–1000	5
<Mn>$_\beta$	34.86	2.76	—	—	1000–1374	5
<Mn>$_\gamma$	44.77	—	—	—	1374–1410	5
<Mn>$_\delta$	47.28	—	—	—	1410–M.P.	5
(N$_2$)	27.87	4.27	—	—	298–2500	4
Glauc.[d]	1717.50	–121.07	70.75	–19.272	298–	2
<PbS>	44.60	16.40	—	—	298–900	4
<S>$_\alpha$	14.98	26.12	—	—	298–369	1
<S>$_\beta$	14.90	29.08	—	—	369–392	1
(S$_2$)	8.54	0.28	–0.79	—	298–2000	5
[S]	22.59	20.92	—	—	392–718	1
<Si>	24.10	2.34	–4.56	—	298–M.P.	5
[Si]	25.61	—	—	—	M.P.–1873	5
<Si$_3$N$_4$>	70.42	98.75	—	—	298–900	5
<SiO$_2$>$_\alpha$	44.60	37.754	–10.018	—	298–844	6

Table A.3A Continued

Heat Capacities: $C_p = a + bT + cT^{-2} + dT^{-0.5}$. Units: J/(mol·K), T is in degrees K, M.P. denotes melting point. $P^0 = 1$ atm (1 bar for References 2 and 6 below). Solid < >, Liquid [ ], Gas ( ).

Substance	a	$b \times 10^3$	$c \times 10^{-5}$	$d \times 10^{-3}$	Temp. Range (K)	Ref.*
<Ti>$_\alpha$	22.09	10.04	—	—	298–1155	4
<Ti>$_\beta$	28.91	—	—	—	1155–1350	5
<Ti>$_\beta$	28.91	—	—	—	1155–1350	5
<TiC>	49.50	3.35	–14.98	—	298–1800	4

*References:

(1) Handbook of Chemistry and Physics, 1989, p. D-44.

(2) Holland, 1990, Table 7, p. 103–104.

(3) Kelley, 1960.

(4) Kubaschewski and Alcock, 1979, Tables C_1 and C_2, p. 336.

(5) Kubaschewski and Evans, 1958, Table C, p. 310.

(6) Robie et al., 1979.

Footnotes

(a) Graphite polymorph

(b) Tremolite: <Ca$_2$Mg$_5$Si$_8$O$_{22}$(OH)$_2$>

(c) Jadeite: <NaAlSi$_2$O$_6$>

(d) Glaucophane: <Na$_2$Mg$_3$Al$_2$(Si$_4$O$_{11}$)$_2$(OH)$_2$>

Table A.3B*

Heat Capacities: $C_p = a + bT + cT^2 + dT^{-0.5} + eT^{-2}$.

Units: J/(mol·K), T is in degrees K, M.P. denotes melting point. $P^0 = 1$ bar. Solid < >, Liquid [ ], Gas ( ).

Substance	a	$b \times 10^3$	$c \times 10^7$	$d \times 10^{-2}$	$e \times 10^{-4}$	Temp. Range (K)
Ab[a]	583.94	–92.852	227.22	–64.242	167.80	298–1500
Arag[b]	81.533	45.673	—	—	–114.05	298–1000
Calc.[c]	99.715	26.920	—	—	–215.76	298–1200
(CO$_2$)	87.820	–2.6442	—	–9.9886	70.641	298–1800
<Cu>	29.764	1.6124	3.411	–1.0067	—	298–1356
(H$_2$)	7.4424	11.707	–13.899	4.1017	–51.041	298–1800
(H$_2$O)	7.368	27.468	–48.117	3.6174	–22.316	298–1800
(O$_2$)	48.318	–0.6913	—	–4.2066	49.923	298–1800
(S$_2$)	47.069	–3.4459	6.9404	–2.3124	–2.2639	298–1800
Serp.[d]	899.60	–144.76	—	–109.32	449.99	298–900
Talc[e]	5653.6	–5271.7	27.291	–769.26	4021.1	298–800
Woll.[f]	111.25	14.373	—	0.1694	–277.79	298–1400

*Robie et al., 1979.

(a) Albite: <NaAlSi$_3$O$_8$>

(b) Aragonite: <CaCO$_3$> polymorph

(c) Calcite: <CaCO$_3$> polymorph

(d) Serpentine: <Mg$_3$Si$_2$O$_5$(OH)$_4$>

(e) Talc: <Mg$_3$Si$_4$O$_{10}$(OH)$_2$>

(f) Wollastonite: <CaSiO$_3$>

Table A.4
Standard Gibbs Free Energy Changes of Selected Reactions:

$$\Delta G_T^0 = A + BT \log(T) + CT \ (\Delta C_p \text{ assumed constant}).$$

Units: J/mol, T is in degrees K. P^0 = 1 atm.

Solid < >, Liquid [ ], Gas ( ).

Reaction	A	B	C	Temp. Range (K)	Ref.[*]
$2<B> + (N_2) = 2<BN>$	−217,590	—	81.18	1200–2300	2
$3<Be> + (N_2) = <Be_3N_2>$	−563,640	—	169.90	298–1000	2
$3<C> + 4[Al] = <Al_4C_3>$	−266,521	—	96.23	932–2000	2
$<C> + 2(H_2) = (CH_4)$	−69,126	51.26	−65.36	298–1200	2
$<C> + 1/2(O_2) = (CO)$	−111,720	—	−87.66	298–2500	2
$<C> + (O_2) = (CO_2)$	−394,170	—	−0.84	298–2000	2
$[Ca] + 2<C> = <CaC_2>$	−57,326	—	−28.45	1123–1963	2
$3<Ca> + (N_2) = <Ca_3N_2>$	−439,360	—	209.20	298–1100	2
$<CaO> + (CO_2) = <CaCO_3>$	−168,420	—	143.94	298–1150	3
$23/6<Cr> + <C> = 1/6<Cr_{23}C_6>$	−68,540	—	−6.44	298–1673	2
$2<Cr> + 3/2(O_2) = <Cr_2O_3>$	−1,120,370	—	259.85	298–2100	3
$2<Cu> + 1/2(O_2) = <Cu_2O>$	−169,470	−16.40	123.44	298–1356	2
$3<Fe> + <C> = <Fe_3C>$	25,940	—	−23.14	298–463	3
$3<Fe> + <C> = <Fe_3C>$	26,700	—	−24.77	463–1115	3
$<Fe> + 1/2(O_2) = <FeO>$	−259,600	—	62.55	298–1642	1
$3<FeCO_3> = <Fe_3O_4>$ $+ 2(CO_2) + (CO)$	220,915	—	−46.94	298–700	3
$1/2(H_2) + 1/2(Cl_2) = (HCl)$	−91,094	4.14	−21.84	298–2100	2
$(H_2) + 1/2(O_2) = (H_2O)$	−239,560	18.75	−9.25	298–2500	2
or alternately	−246,460	—	54.82	298–2500	2
$2(H_2) + (S_2) = 2(H_2S)$	−180,600	—	98.79	298–1800	2
$[Mn] + 1/2(S_2) = <MnS>$	−288,770	—	78.92	1517–1803	2
$4<Mo> + (N_2) = 2(Mo_2N)$	−143,940	−38.50	242.30	298–1300	2
$2(N_2) + 3<Si> = <Si_3N_4>$	−753,190	—	336.43	298–1680	3
$2(N_2) + 3[Si] = <Si_3N_4>$	−892,950	—	419.28	1680–1800	3
$<Nb> + <C> = <NbC>$	−130,140	—	1.67	1180–1370	2
$<Ni> + 1/2(O_2) = <NiO>$	−244,580	—	98.54	298–1725	3
$[Pb] + 1/2(O_2) = <PbO>$	−229,930	−33.68	209.64	600–1150	3
$(S_2) + 2(O_2) = 2(SO_2)$	−724,910	—	144.90	298–2000	2
$<Si> + (O_2) = <SiO_2>$	−881,235	−12.55	218.51	298–1700	3
$<Si> + <SiO_2> = 2(SiO)$	697,540	53.98	−518.45	298–1700	3
$2<Ti>_\alpha + (N_2) = 2<TiN>$	−671,600	—	185.80	298–1155	2
$2<Ti>_\beta + (N_2) = 2<TiN>$	−676,620	—	190.20	1155–1500	2
$<W> + <C> = <WC>$	−37,656	—	1.674	298–2000	2

[*]References:

(1) Gaskell, 1981, Table A-1, p. 585.

(2) Kubaschewski and Alcock, 1979, Table E, p. 378.

(3) Kubaschewski and Evans, 1958, Table E, p. 336.

Table A.5*
Vapor Pressures Over Pure Condensed Phases.
$\log_{10}(P)$ (mm Hg) $= AT^{-1} + B\log_{10}(T) + CT + D$.
T is in degrees K. Solid < >, Liquid [ ].

Substance	A	B	$C \times 10^2$	D	Temp. Range (K)
<Cu>	–17,770	–0.86	—	12.29	298–1356
[Cu]	–17,520	–1.21	—	13.21	1356–2843
<Zn>	–6850	–0.755	—	11.24	298–693
[Zn]	–6620	–1.255	—	12.34	693–1180

*Kubaschewski and Alcock, 1979, Table D, p. 358.

Table A.6*
Activity of Carbon in Austenite Between 800–1200°C.

w/o C	$a_C^\dagger$	w/o C	$a_C^\dagger$
0.05	0.00236	1.1	0.0728
0.10	0.00479	1.2	0.0822
0.20	0.00992	1.3	0.0917
0.30	0.01537	1.4	0.1022
0.40	0.0211	1.5	0.1130
0.50	0.0273	1.6	0.1245
0.60	0.0338	1.7	0.1369
0.70	0.0407	1.8	0.1495
0.80	0.0480	1.9	0.1632
0.90	0.0559	2.0	0.1774
1.00	0.0640	—	—

* Darken and Gurry, 1953, Table 16-4, p. 405-406.

† Standard state is such that $\lim_{X_C \to 0} (a_C / X_C) = 1$. The activity of carbon relative to graphite is obtained from the conversion a_C (graphite) $= a_C$ (w/o C in austenite)/a_c(w/o C in austenite at saturation) at the same temperature.

APPENDIX B

TABLES OF SELECTED PHYSICAL CONSTANTS

Table B.1
Density, Molecular Mass, Volume Thermal Expansion and Isothermal
Compressibility Coefficients
Solid < >, Liquid [ ]

Substance	ρ gm/cm^3	M gm/(gm·mol)	α K^{-1}	β atm^{-1}
<Ag>[a]	10.49	107.88	19.68 x 10^{-6}	—
<Au>[a]	19.3	197	—	—
[Au][d]	17.0	197	—	—
<C>[a,1]	2.25	12.01	—	—
<Ca$_\alpha$>[b,1]	1.55	40.08	—	—
<Cu>[a,c]	8.96	63.54	16.50 x 10^{-6}	7.60 x 10^{-6}
<Fe>[a]	7.85	55.85	—	—
<Fe$_3$C>[d]	7.40	179.40	—	—
H[e]	—	1.01	—	—
K[e]	—	39.10	—	—
Li[e]	0.53	6.941	—	—
O[e]	—	15.99	—	—
Pb[e]	—	207.2	—	—
S[e]	—	32.06	—	—
Sb[e]	6.68	121.75	—	—
<Si>[a]	2.35	28.09	—	—
[Si][b]	2.57	28.09	—	—
Ti[e]	—	47.90	—	—

References:

  (a) Metals Handbook, Desk Edition, 1985, Chapter 1, p. 44.

     (1) <C> = Graphite.

  (b) Handbook of Chemistry and Physics, 1989, 70th edition, p. B-11.

     (1) The density, ρ, of <Ca> at 293 K.

  (c) Darken and Gurry, 1953, p. 498 for β.

  (d) Handbook of Chemistry and Physics, 1948, 30th edition, p. 368.

  (e) Periodic Table of the Elements, 1979, Sargent-Welch Scientific Co.

Table B.2
Standard Molar Volumes, Volume Thermal Expansion
and Isothermal Compressibility Coefficients
$P^0 = 1$ bar, $T = 298$ K.
Solid < >, Liquid [ ], Gas ( ).

Substance	V cm^3/mol	α K^{-1}	β atm^{-1}	Ref.*
Graphite $<C>$	5.30	$28{,}301.89 \times 10^{-9}$	2641.51×10^{-9}	1
Lawsonite[a]	101.32	2.47×10^{-5}	8.90×10^{-7}	1
Aragonite $<CaCO_3>$	34.15	—	—	2
Tremolite[b]	272.70	30.99×10^{-6}	1.32×10^{-6}	1
$<CaSiO_3>$[c]	39.93	2.404×10^{-5}	901.58×10^{-9}	1
$[H_2O]$	18.069	—	—	2
Spinel[d]	39.78	25.89×10^{-6}	477.63×10^{-9}	1
Pyrope Garnet[e]	113.18	2.63×10^{-5}	5.57×10^{-7}	1
(O_2)	24,789.20	—	—	2

* References:

(1) Holland, 1990, Table 7, p. 103.

(2) Robie et al., 1979.

Footnotes:

(a) Lawsonite: $<CaAl_2Si_2O_7(OH)_2{\cdot}H_2O>$

(b) Tremolite:$<Ca_2Mg_5Si_8O_{22}(OH)_2>$

(c) Wollastonite: $<CaSiO_3>$

(d) Spinel: $<MgAl_2O_4>$

(e) Pyrope Garnet: $<Mg_3Al_2Si_3O_{12}>$

APPENDIX C

ELECTROMOTIVE FORCE SERIES

Table C.1*
Half Cell Reactions and Standard Potentials
(Engineering Sign Convention)

Half Cell Anode Reactions[†]	E^0 (Potential in Volts)[†]
$Au \rightarrow Au^{3+} + 3e^-$	+1.50
$Ag \rightarrow Ag^{1+} + 1e^-$	+0.80
$Cu \rightarrow Cu^{2+} + 2e^-$	+0.34
$Sn \rightarrow Sn^{2+} + 2e^-$	−0.14
$Fe \rightarrow Fe^{2+} + 2e^-$	−0.44
$Zn \rightarrow Zn^{2+} + 2e^-$	−0.76
$Al \rightarrow Al^{3+} + 3e^-$	−1.66
$Li \rightarrow Li^{1+} + 1e^-$	−3.05

* Askeland, D.R., 1989, Table 20-1, p. 784.

† Potential relative to hydrogen electrode. For cathode reaction, reverse the anode reaction and change the sign on E^0.

PROPERTIES AT VAPOR-LIQUID EQUILIBRIUM

Table D.1*
Saturated Water Vapor-Liquid Equilibrium

| Temp. | Pressure | Enthalpy kJ/kg | | Entropy kJ/(kg·K) | |
| | | Liquid | Vapor | Liquid | Vapor |
°C	Bar	h_l	h_g	s_l	s_g
25	0.03169	104.89	2547.2	0.3674	8.5580
65	0.2503	272.06	2618.3	0.8935	7.8310
100	1.014	419.04	2676.1	1.3069	7.3549

* Wark, K., 1983, Table A-12M, p. 799.

Table D.2*
Potassium Vapor-Liquid Equilibrium

| Temp. K | Bar | Enthalpy kJ/kg | | Entropy kJ/(kg·K) | | Volume L/kg | | Internal Energy kJ/kg |
| | | Liquid | Vapor | Liquid | Vapor | Liquid | Vapor | |
		h_l	h_g	s_l	s_g	v_l	v_g	$u_g - u_l$
900	0.2510	731.0	2739.7	2.6874	4.9175	1.438	7180	1829
950	0.4470	771.4	2750.4	2.7313	4.8129	1.462	4204	1791
1000	0.7530	812.6	2760.4	2.7732	4.7196	1.493	2592	1753

* Wark, K., 1983, Table A-23m, p. 817.

APPENDIX E

FIGURES

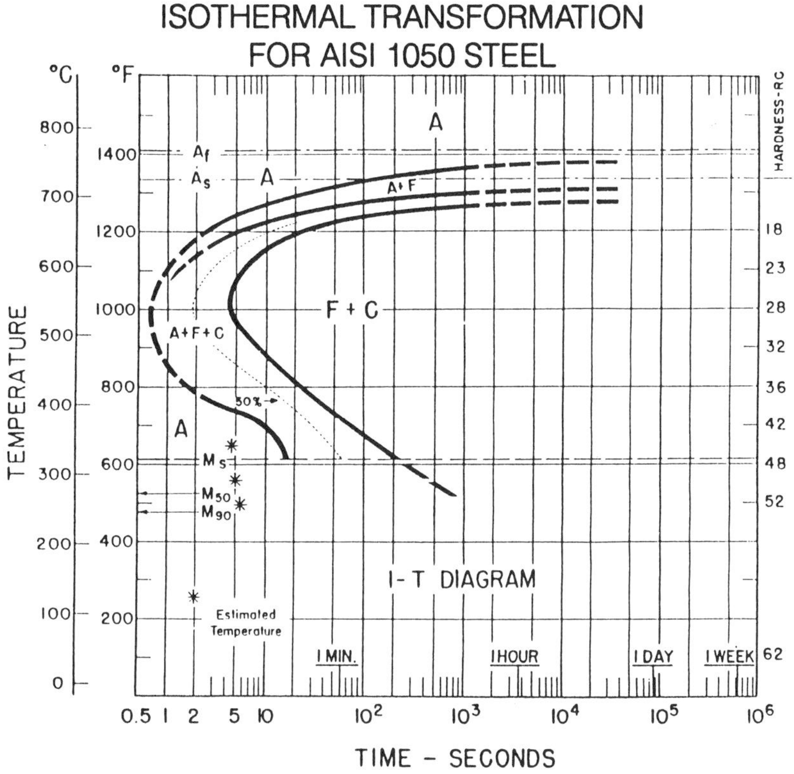

Figure E.1 **Isothermal** Transformation (I-T) diagram for AISI 1050 Steel. (From the Atlas of Isothermal Transformation and Cooling Transformation Diagrams, 1977, p. 15. Reprinted by permission of ASM International, Metals Park, Ohio.)

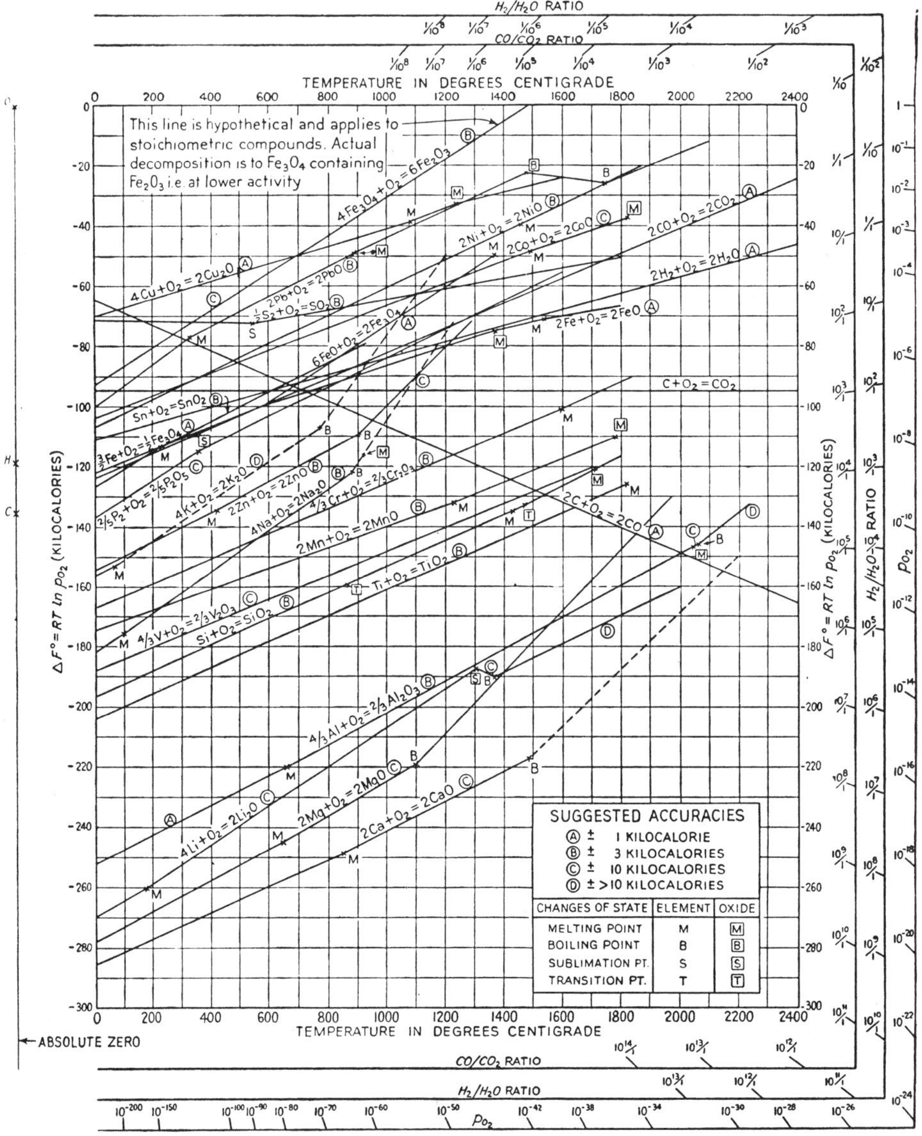

Figure E.2 The standard free energy of formation of metal oxides as a function of temperature. (From L.S. Darken and R.W. Gurry, 1953, Physical Chemistry of Metals, Fig. 14-4. Reprinted by permission of McGraw-Hill, Inc., New York.)

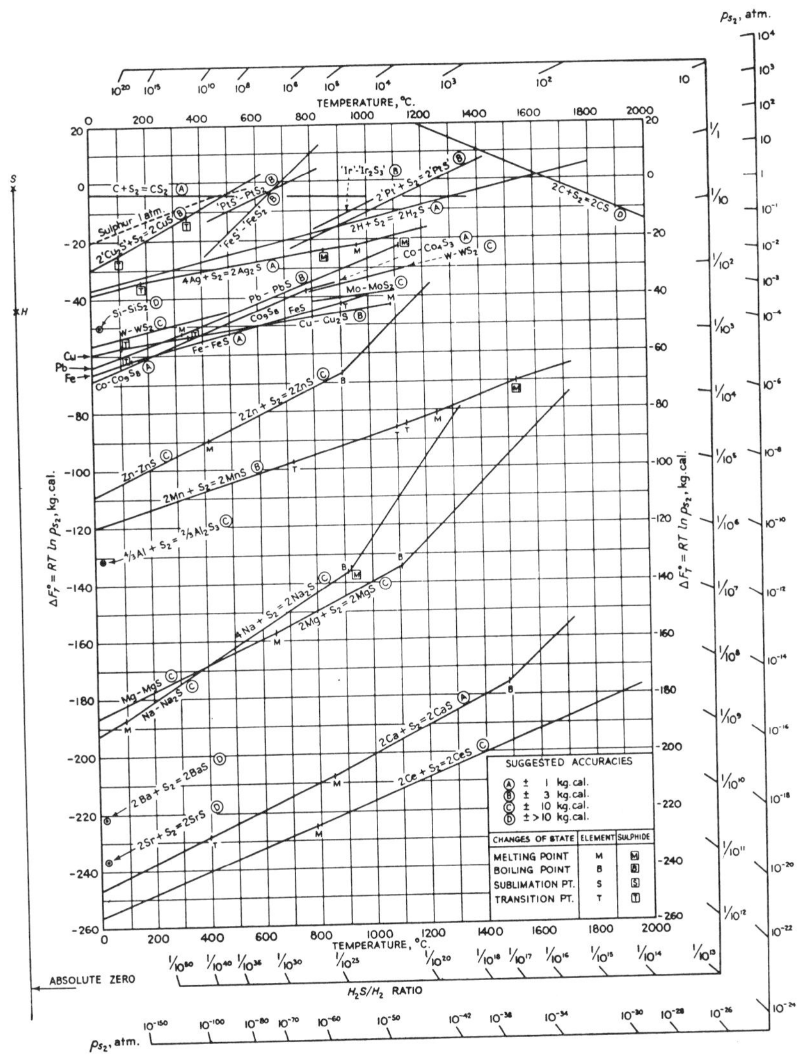

Figure E.3 The standard free energy of formation of metal sulfides as a function of temperature. (From L.S. Darken and R.W. Gurry, 1953, Physical Chemistry of Metals, Fig. 14-9. Reprinted by permission of McGraw-Hill, Inc., New York.)

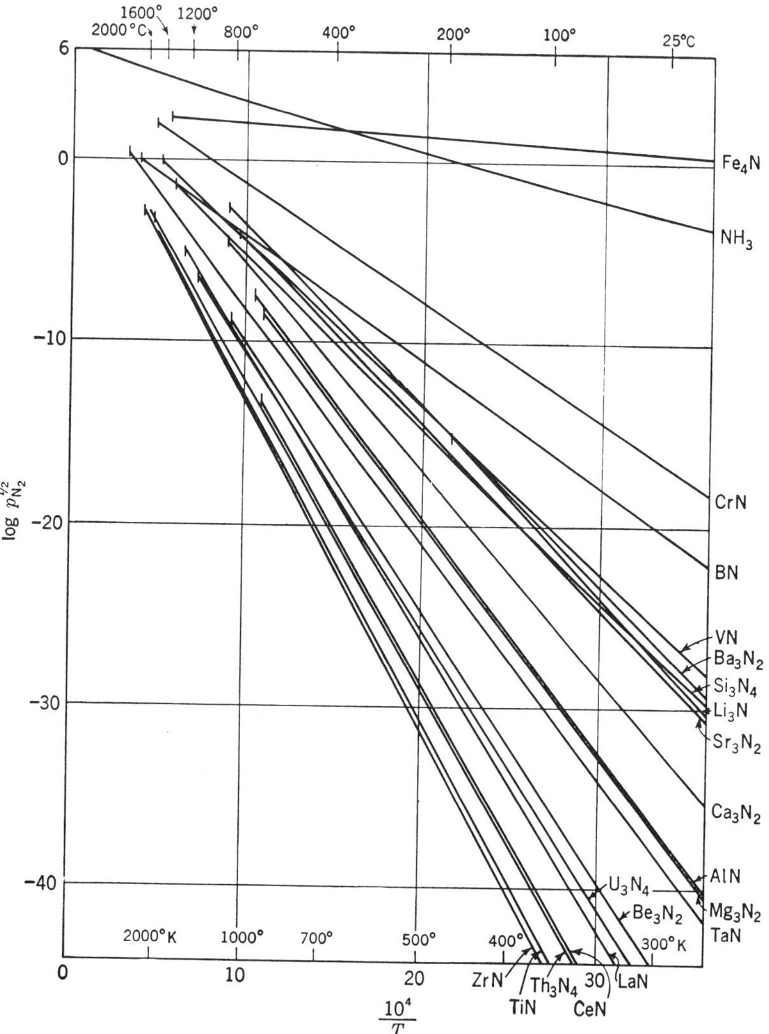

Figure E.4 Log $P_{N_2}^{1/2}$ vs. $10^4/T$ for select metal-nitride systems. (From L.S. Darken and R.W. Gurry, 1953, *Physical Chemistry of Metals*, Fig. 14-11. Reprinted by permission of McGraw-Hill, Inc., New York.)

E^0 AND $-\Delta F^0$ $(-\Delta G^0)$ FOR SELECT CHLORIDES AS A FUNCTION OF TEMPERATURE

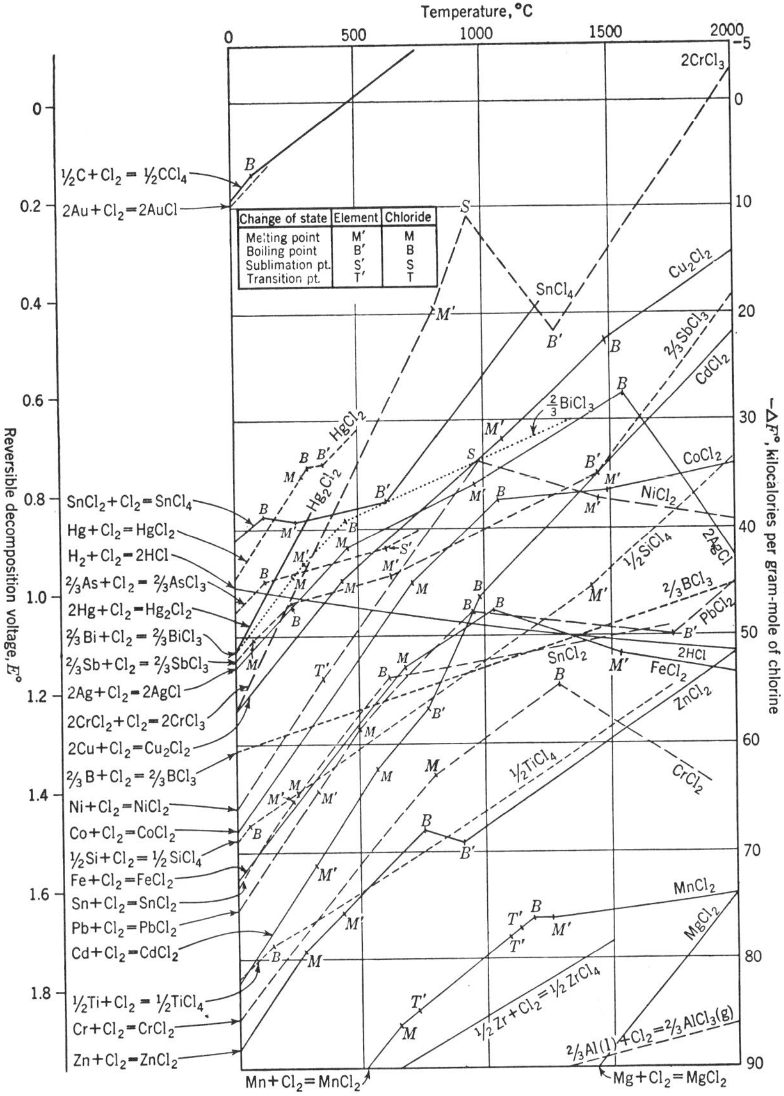

Figure E.5 The standard free energy of formation and reversible decomposition voltage of select chlorides as a function of temperature. (From L.S. Darken and R.W. Gurry, 1953, Physical Chemistry of Metals, Fig. 14-12. Reprinted by permission of McGraw-Hill, Inc., New York.)

ALUMINUM-SILICON EUTECTIC PHASE DIAGRAM

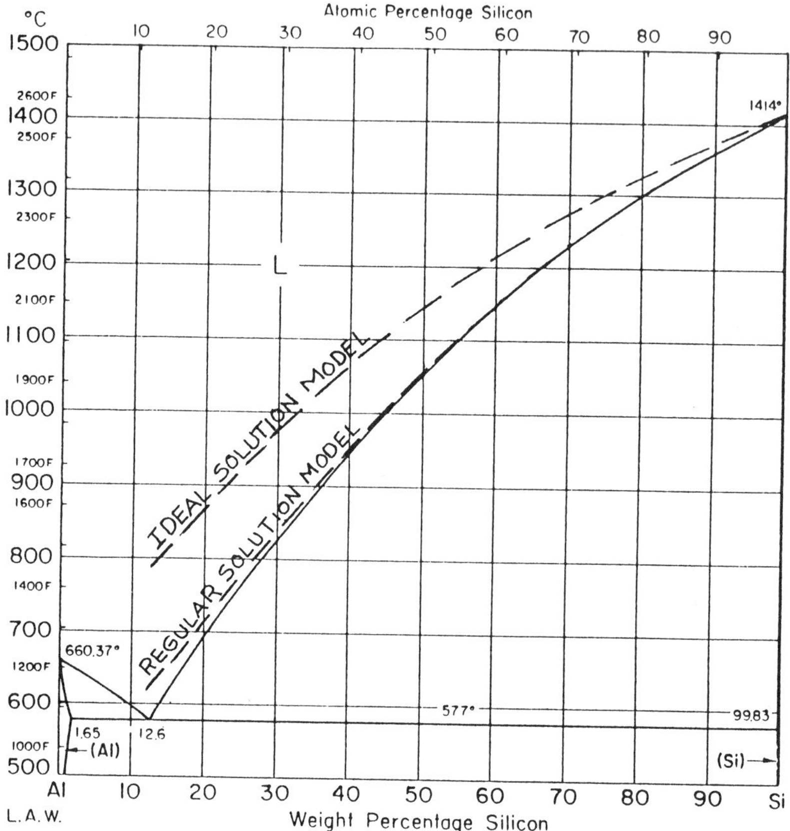

Figure E.6 Al-Si Eutectic Phase Diagram: No solid solubility. (From the Metals Handbook, 8th edition, v. 8, Metallography, Structures, and Phase Diagrams, 1973, p. 263. Reprinted by permission of ASM International, Metals Park, Ohio.)

SILVER-COPPER EUTECTIC PHASE DIAGRAM

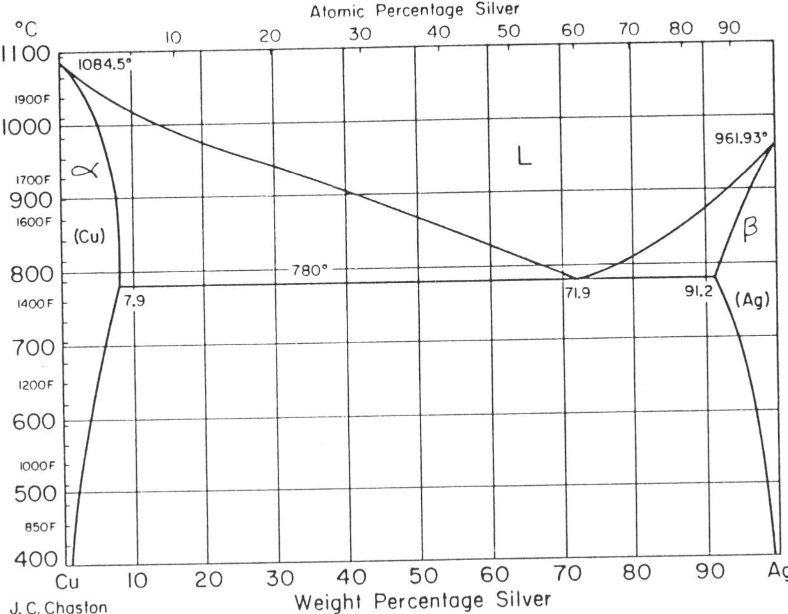

Figure E.7 Ag-Cu Eutectic Phase Diagram: Terminal Solid Solubility. (From the Metals Handbook, 8th edition, v. 8, Metallography, Structures, and Phase Diagrams, 1973, p. 253. Reprinted by permission of ASM International, Metals Park, Ohio.)

BISMUTH-LEAD PHASE DIAGRAM

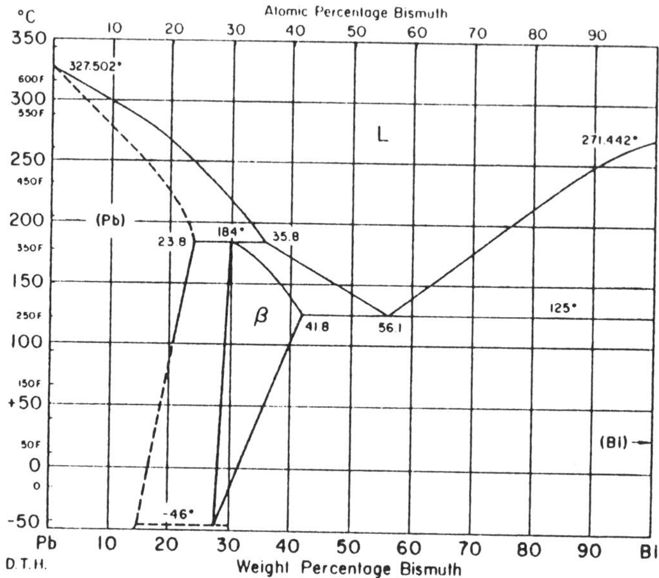

Figure E.8 Bi-Pb Phase Diagram. (From the Metals Handbook, 8th edition, v. 8, Metallography, Structures, and Phase Diagrams, 1973, p. 273. Reprinted by permission of ASM International, Metals Park, Ohio.)

LEAD-ANTIMONY PHASE DIAGRAM

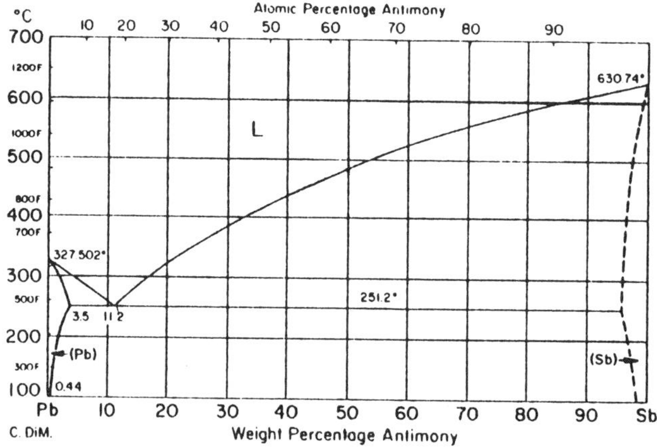

Figure E.9 Pb-Sb Phase Diagram. (From the Metals Handbook, 8th edition, v. 8, Metallography, Structures, and Phase Diagrams, 1973, p. 329. Reprinted by permission of ASM International, Metals Park, Ohio.)

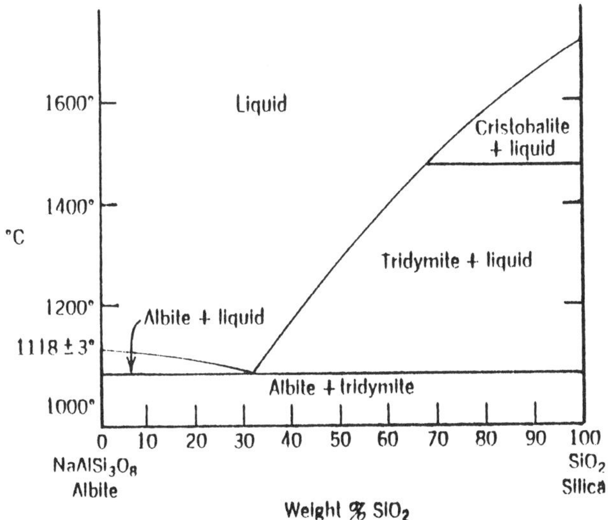

Figure E.10 NaAlSi3O8-SiO2 eutectic phase diagram: No solid solubility. (From C. Klein and C.S. Hurlbut, Jr., 1985, Manual of Mineralogy, Fig. 12.5. Reprinted by permission of John Wiley and Sons, Inc., Copyright © 1985.)

DIAMOND-GRAPHITE UNIVARIANT PHASE DIAGRAM

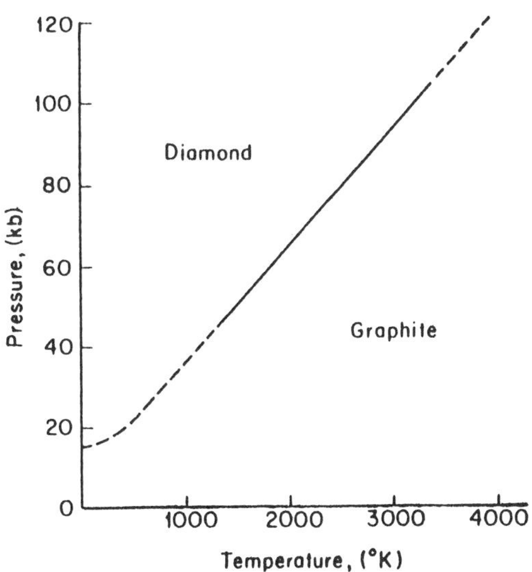

Figure E.11 Diamond-graphite univariant P-T phase diagram. (From E.G. Ehlers, 1972, The Interpretation of Geological Phase Diagrams, Fig. 93. Reprinted by permission of W.H. Freeman and Company.)

REFERENCES FOR THERMODYNAMIC DATA

Anyalebechi, P.N., Talbot, D.E.J., and Granger, D.A., 1988, *The solubility of hydrogen in liquid binary Al-Li alloys*, Met. Trans. B, v. 19B, no. 2, p. 227–232.

Askeland, D.R., 1989, *The Science and Engineering of Materials*, 2nd edition, PWS-Kent Publishing Co., Boston, Mass., 876 p.

Berman, R.G., 1988, *Internally-consistent thermodynamic data for minerals in the system $Na_2O-K_2O-CaO-MgO-FeO-Fe_2O_3-Al_2O_3-SiO_2-TiO_2-H_2O-CO_2$*, Jour. Petrol., v. 29, part 2, p. 445–522.

Ellingham, H.J.T., 1944, *Reproducibility of oxides and sulfides in metallurgical processes*, Jour. Soc. Chem. Ind., v. 63, p. 125.

Handbook of Chemistry and Physics, 1948, 30th edition, Hodgman, C.D., editor in chief, Chemical Rubber Publishing Co., Cleveland, Ohio, 2686 p.

Handbook of Chemistry and Physics, 1989, 70th edition (1989–1990), Weast, R.C., senior editor, CRC Press, Inc., Boca Raton, Florida.

Holland, T.J.B., 1990, *An enlarged and updated internally consistent thermodynamic data set with uncertainties and correlations: the system $K_2O-Na_2O-CaO-MgO-MnO-FeO-Fe_2O_3-Al_2O_3-TiO_2-SiO_2-C-H_2-O_2$*, Jour. Met. Petrol., v. 8, p. 89–124.

Iwai, T., Takahashi, I., and Handa, M., 1986, *Gibbs free energies of formation of Molybdenum Carbide and Tungsten Carbide from 1173 to 1573 K*, Met. Trans. A, v. 17A, p. 2031–34.

JANAF Thermochemical Tables, 2nd ed., 1971, Stull, D.R. and Prophet, H., project directors, U.S. Dept. of Commerce, National Bureau of Standards NSRDS-NBS 37, Washington, D.C.

Johnson, D.L., 1964, *Phase equilibria-free energy relationships in liquid sodium systems*, NAA-SR-8381, Atomics International (For AEC), p. 1–39.

Kelly, K.K., 1960, *High-Temperature Heat-Content, Heat Capacity, and Entropy Data for the Elements and Inorganic Compounds*, U.S. Bur. Mines Bull. 584, 232 p.

Kubaschewski, O., and Evans, E.L., 1958, *Metallurgical Thermochemistry*, 3rd edition, Pergamon Press, London, England, 426 p.

Kubaschewski, O., and Alcock, C.B., 1979, *Metallurgical Thermochemistry*, 5th edition, Pergamon Press, Oxford, England, 449 p.

McQuillan, A.D., 1950, *An experimental and thermodynamic investigation of the hydrogen-titanium system*, Proc. Roy. Soc. London, v. 204, p. 309–23.

Metals Handbook, Desk Edition, 1985, Gall, T.L., project director, The American Society for Metals, 2nd printing, Metals Park, Ohio.

Metals Handbook: Corrosion, 1987, v. 13, The American Society for Metals, Metals Park, Ohio, 1415 p.

Periodic Table of the Elements, 1979, Catalog No. S-18806, Sargent-Welch Scientific Co., 7300 North Linder Ave., Skokie, Illinois 60077.

Robie, R.A., Hemingway, B.S., and Fisher, J.R., 1979, *Thermodynamic Properties of Minerals and Related Substances at 298.15 K and 1 Bar (10^5 Pascals) Pressure and at Higher Temperatures*, U.S. Geol. Survey Bull. 1452, 456 p.

Schmalz, R.F., 1959, *A note on the system Fe_2O_3-H_2O*: Jour. Geophys. Research, v. 64, p. 575–579.

Simensen, C.J., 1989, *Comments on the solubility of carbon in molten aluminum*, Met. Trans. A, v. 20A, p. 191.

Smith, W.F., 1986, *Principles of Materials Science and Engineering*, McGraw Hill, New York, NY/St. Louis, MO., 777 p.

Wark, K., 1983, *Thermodynamics*, 4th edition, McGraw-Hill, New York, NY/St. Louis, MO., 896 p.

Weaver, A.B., Johnson, D.L., and St. Pierre, G.R., 1977, *Thermodynamic computation of phase equilibria in the sodium-carbon-oxygen system*, Met. Trans. A, v. 8A, p. 603–607.

Wicks, C.E., and Block, F.E., 1963, *Thermodynamic properties of 65 Elements—Their Oxides, Halides, Carbides, and Nitrides*, U.S. Bur. Mines Bull. 605, Albany Metallurgical Research Center, U.S. Bureau of Mines, Albany, Oregon, 146 p.

ADDITIONAL REFERENCES

Bowen, N.L., 1913, *The melting phenomena of the plagioclase feldspars*, Amer. Jour. Sci., 4th series, v. 34, p. 577–99.

Boyd, F.R., and England, J.L., 1960, *The quartz-coesite transition*, Jour. Geophys. Res., v. 65, p. 749–56.

Bundy, F.P., Bovenkerk, H.P., Strong, H.M., and Wentorf, R.H. Jr., 1961, *Diamond-graphite equilibrium line from growth and graphitization of diamond*, Jour. Chem. Physics, v. 35, p. 383–391.

Curzon, F.L., and Ahlborn, B., 1975, *Efficiency of a Carnot engine at maximum power output*, Am. Jour. Phys., v. 48, p. 22–24.

Darken, L.S., and Gurry, R.W., 1953, *Physical Chemistry of Metals*, McGraw-Hill, New York, NY/St. Louis, MO., 535 p.

DeHoff, R.T., 1993, *Thermodynamics in Materials Science*, McGraw-Hill, New York, NY/St. Louis, MO., 532 p.

Ehlers, E.G., 1972, *The Interpretation of Geological Phase Diagrams*, W.H. Freeman and Company, San Francisco, Calif., 280 p.

Fine, H.A., and Geiger, G.H., 1979, *Handbook on Material and Energy Balance Calculations in Metallurgical Processes*, The Metallurgical Society of AIME, Warrendale, Pennsylvania 572 p.

Gaskell, D.R., 1981, *Introduction to Metallurgical Thermodynamics*, 20th edition, McGraw-Hill, New York, NY/St. Louis, MO., 611 p.

Gokhale, A.A., and Johnson, D.L., 1982, *Recomputation in the sodium-carbon-oxygen system: effect of oxygen*, Met. Trans. A, v. 13A, p. 1101–1102.

Hamill, W.H., Williams, R.R. Jr., and MacKay, C., 1966, *Principles of Physical Chemistry*, 2nd edition, Prentice- Hall, Inc., Englewood Cliffs, New Jersey, 576 p.

Kane, R.D. and Chakachery, E.A., 1992, H_2 *effects on Ti-Al and C/C composites*, ASM News.

Klein, C., and Hurlbut, C.S. Jr., 1985, *Manual of Mineralogy*, 20th edition, John Wiley and Sons, New York, NY/Toronto, Ontario, 596 p.

Lupis, C.H.P., 1983, *Chemical Thermodynamics of Materials*, Elsevier Science Publishing Co., New York, NY, 581 p.

NBS Special Publication 330, 1972, *The International System of Units (SI)*, U.S. Government Printing Office, Washington, D.C. 20402.

Polzin, M.H., 1951, *Performance evaluation of a magnesium alloy truck wheel*, Proc. Soc. Exp. Stress Analysis, v. 11, no. 1, p. 1–16.

Protter, P.H. and Morrey, C.B. Jr., 1970, *College Calculus with Analytic Geometry*, 2nd edition, Addison-Wesley, Reading, Mass./Menlo Park, Calif., 900 p.

Schairer, J.F. and Bowen, N.L., 1956, *The system Na_2O-Al_2O_3-SiO_2*, Amer. Jour. Sci., v. 254, p. 129–195.

Stracher, G.B., and Johnson, D.L., 1990, *Thermodynamic loops: theoretical applications to solid-solid mineral transformations and phase equilibria*, The Compass, v. 66, no. 4, p. 185–188.

Swalin, R.A., 1964, *Thermodynamics of Solids*, John Wiley and Sons, New York, NY/ Toronto, Ontario, 343 p.

Thomsen, J.S., 1962, A *restatement of the zeroth law of thermodynamics*, Amer. Jour. Phys., v. 30, p. 294.

Upadhyaya, G.S. and Dube, R.K., 1977, *Problems in Metallurgical Thermodynamics and Kinetics*, Pergamon Press, Elmsford, New York, 252 p.

Van Vlack, L.H., 1985, E*lements of Materials Science and Engineering*, 5th edition, Addison-Wesley, Reading, Mass./Menlo Park, Calif., 633 p.

Wilkinson, W.D., and Murphy, W.F., 1958, *Nuclear Reactor Metallurgy*, D. Van Nostrand Co., Inc., Princeton, New Jersey, 382 p.

Wu, H.C., 1992, T*hermodynamic calculation of partial phase diagram of Al-Si alloy at high pressure*, Jour. of Mtrls. Sci. Ltrs., v. 11, p. 1–5.

Zemansky, M.W. and Van Ness, H.C., 1966, *Basic Engineering Thermodynamics*, McGraw-Hill, New York, NY/St. Louis, MO., 380 p.